High Voltage Equipment Testing
（高电压设备测试）

何发武　曾庆洪　冯文艳 ◎ 著

西南交通大学出版社
·成　都·

图书在版编目（CIP）数据

高电压设备测试 = High Voltage Equipment Testing：英文 / 何发武，曾庆洪，冯文艳著. -- 成都：西南交通大学出版社，2023.12
ISBN 978-7-5643-9682-4

Ⅰ. ①高… Ⅱ. ①何… ②曾… ③冯… Ⅲ. ①高压电器－测试－高等职业教育－教材－英文 Ⅳ. ①TM510.6

中国国家版本馆 CIP 数据核字（2023）第 246648 号

High Voltage Equipment Testing
（高电压设备测试）

何发武　曾庆洪　冯文艳　著

责任编辑	李　伟
封面设计	吴　兵

出版发行	西南交通大学出版社 （四川省成都市金牛区二环路北一段 111 号 西南交通大学创新大厦 21 楼）
邮政编码	610031
营销部电话	028-87600564　028-87600533
网址	http://www.xnjdcbs.com
印刷	四川森林印务有限责任公司

成品尺寸	185 mm × 260 mm
印张	17.5
字数	564 千
版次	2023 年 12 月第 1 版
印次	2023 年 12 月第 1 次
定价	65.00 元
书号	ISBN 978-7-5643-9682-4

课件咨询电话：028-81435775
图书如有印装质量问题　本社负责退换
版权所有　盗版必究　举报电话：028-87600562

前言

在"一带一路"倡议和"中国铁路走出去"战略背景下，中国轨道交通海外建设项目高速发展，海外铁路建设和运营需要大量熟练掌握轨道交通专业英语的技能型人才。

本书以高电压绝缘试验与特性试验为核心图谱，从高电压试验概述、高电压绝缘试验、高电压特性试验和高电压新技术等方面介绍了高电压设备测试的核心内容，满足高电压电气试验工的岗位技能培训要求，也适用于高压变电工、配电工、检修工技能培训。

本书为英文专著，配套有中文数字书稿及丰富的视频资源，包含铁道供电国家资源库，"高电压设备测试"国家在线精品课程中、英文版双语资源。中文课程服务中国高铁供电，英文双语资源助力"中国走出去"战略，承接国家国际发展合作署"一带一路"国家轨道交通建设项目、商务部发展中国家铁路供电技术项目等援外培训，并应用到埃塞俄比亚至吉布提（亚吉）铁路、埃及开罗LRT铁路、印尼雅万高速铁路等项目培训中，资源丰富，实用性强，适合海内外培训。

本书由广州铁路职业技术学院何发武等编写，其中广州地铁集团有限公司曾庆洪编写项目一工单一、工单二，昆明铁道职业技术学院冯文艳编写项目四，广州铁路职业技术学院邓婧编写项目一工单三，广东技术师范大学谢月编写项目三工单十，其余内容由何发武完成并进行全书统稿。何发武和冯文艳完成全书中英文翻译对照工作。

本书是广州铁路职业技术学院"双高计划"资助项目。作者在写作过程中，多次到广铁集团供电段、广州地铁等地进行项目调研；赵灵龙、易江平、罗建群、骆世忠等现场专家对本书的编写提出了宝贵意见；广州铁路职业技术学院电气化铁道技术团队全体成员对本书的编写给予了无私帮助，同时冼明珍、杨波琳、何安洋、李雅、何凤华、何月华、何发文、李凤、何发贤等同志也给予了大力支持，在此一并表示衷心感谢。

由于作者水平有限，书中难免有疏漏之处，希望读者多多指正。

作 者

2023年11月于广州

《高电压设备测试》中文版教材

Contents

Project 1　Overview of High Voltage Testing ·· 001

　　Worksheet 1　High Voltage Safety ·· 001
　　Worksheet 2　High Voltage Insulation Tools ·· 008
　　Worksheet 3　High Voltage Testing and Instruments ·· 015

Project 2　High Voltage Insulation Test ·· 025

　　Worksheet 1　Insulation Resistance Test ··· 025
　　Worksheet 2　Leakage Current and DC Withstand Voltage Test ····························· 033
　　Worksheet 3　Dielectric Loss Test ·· 042
　　Worksheet 4　AC Voltage Withstand and Series Resonance Withstand Test ·········· 051
　　Worksheet 5　Partial Discharge Test ·· 071

Project 3　High Voltage Characteristic Test ·· 085

　　Worksheet 1　DC Resistance Test ·· 085
　　Worksheet 2　Loop Resistance Test ·· 096
　　Worksheet 3　Transformer Ratio and Grouping Measurement ································ 107
　　Worksheet 4　Transformer Winding Deformation Test ·· 117
　　Worksheet 5　Transformer Oil Chromatographic Analysis ······································ 143
　　Worksheet 6　Insulating Oil Dielectric Strength Test ··· 155
　　Worksheet 7　Switch Characteristic Test ··· 167
　　Worksheet 8　Cable Fault Comprehensive Test ··· 180
　　Worksheet 9　Insulating shoe and insulating glove test ·· 211
　　Worksheet 10　High Voltage Nuclear Phase Test ·· 224
　　Worksheet 11　Grounding Resistance Measurement of Ground Grid ····················· 230

Project 4　High Voltage New Technology ·· 242

　　Worksheet 1　Hot Washing ··· 242
　　Worksheet 2　High Voltage Live Detection ··· 253

Appendix　Training answer ··· 265

Bibliography ·· 273

Project 1

Overview of High Voltage Testing

Worksheet 1　High Voltage Safety

Module 1　Operating Worksheet: High Voltage Safety

(Ⅰ) Safety Tool Inspection	(Ⅱ) Safety Regulations
(1) Check the insulating tools and instruments are within the validity period. (2) Check whether the surface of the tool is broken, burst, etc.	(1) Hang a sign with "Stop, High Voltage Danger!" (2) Maintain a sufficient safe distance
(Ⅲ) Test Preparation	(Ⅳ) Test Steps
(1) Test personnel "two-wearing and three-carrying". (2) Set up barriers to prevent extraneous personnel from entering the test area. (3) Mark the cables before disconnection for proper identification, and conduct inspections after disconnection	(1) Check the power before operation, hang the ground wire. (2) Check Protective devices and test instruments inspection. (3) Check Normal high-voltage test operation
(Ⅴ) Notes	(Ⅵ) Site Cleanup
(1) Use safety belts to prevent falling from height. (2) Prevent electric shock injuries from induction and high-voltage contact	(1) Remove test cables and temporary grounding wires, clean up tools and debris at the site. (2) Restore the test cables to their original condition. (3) Remove protective barriers and signs
(Ⅶ) Validity period of tools	(Ⅷ) Digital Resources
(1) High-voltage tools must be used within their validity period. (2) Tools must be inspected according to the test cycle	**Electrical Tools and Appliances Safety Performance Testing**

Module 2 Follow me

High-voltage testing is a high-risk task in electrical testing. During the process of high-voltage testing, it is necessary to comply with relevant national safety regulations to ensure personal safety and the safety of the power system. The main standard for high-voltage testing is referenced from the "Standard for Electrical Equipment Installation Engineering-Electrical Equipment Handover Testing" (GB 50150–2016).

Ⅰ. Safety Distance

The safety distance refers to the minimum air gap maintained between the operator and the live object to ensure personal and equipment safety. The size of the safety distance varies depending on the voltage level, equipment type, installation method, and weather conditions. According to the current safety regulations of the State Grid power plant and substation, the safety distances for live equipment are 0.7 m for 10 kV, 1 m for 35 kV, 1.5 m for 110 kV, 3 m for 220 kV, 4 m for 330 kV, and 5 m for 500 kV.

The safety distance for operating non-power-off equipment in the State Grid's 10 kV distribution station is shown in Figure 1-1.

Figure 1-1 Safety distance for live operation of 10 kV distribution equipment in the state grid

When the distance between the live object and the human body, equipment, and live wires during testing does not meet the requirements, temporary barriers, insulating shields, insulating mats, etc., must be installed for isolation. If sparks or discharge sounds occur during the testing, it indicates that the distance is not sufficient. Stop the test immediately, adjust the distance, wipe the insulation surface, and test again.

II. Safety Regulations

High-voltage test safety operating procedures: power outage, power testing, grounding, hanging signage, each step must be performed in strict accordance with the order. One person in operation, one person in guard.

According to the content of the work permission, the power line and equipment should be confirmed to cut off from the power. The corresponding voltage level electroscope should be used in the electricity test, and the electricity should be tested item by item at the installation of the grounding wire or the closing of the grounding switch. When grounding, after confirming there is no voltage, the high-voltage lines and equipment should be grounded and short-circuited immediately, and all lines that may be reversely powered should be grounded.

The climate conditions for high voltage insulation testing should not be lower than +5 °C, and the relative humidity of the air should generally not exceed 80%. The testing sequence of "non-destructive testing before destructive testing" should be followed. If the weather humidity is too high, the insulation surface may condense or form a water film, which will reduce the surface insulation resistance, greatly increase the surface leakage current, and also cause distortion of the electric field distribution on the insulation surface, resulting in corona discharge, affecting the measurement results, and requiring testing to be rescheduled.

Before and after the testing, establish safety monitoring and protection systems. The testing personnel should wear necessary protective equipment before the testing ("two-wearing and three-carrying": wearing working uniform, wearing insulating boots, carrying safety helmets, carrying insulating gloves, and carrying voltage detectors). The testing site should set up fences and hang "Stop, High Voltage Danger" signs facing outward. For operations above 2 m, adequate fall protection measures should be taken. Personnel and equipment should maintain a safe testing distance, as shown in Figure 1-2.

Figure 1-2　Substation test electrical grounding

Module 3　Workshop

Ⅰ. Work specification

The safety level of the workers must not be lower than level three. Workers should fasten their seat belts, wear safety helmets, and take anti-slip measures. Work can only be carried out in clear weather conditions, starting from 9:00 AM. One supervisor can only oversee one group of workers. Grounding workers must wear safety helmets. The tools carried by the workers must be placed in tool bags, and the bags must be securely fastened to prevent tools from falling and injuring ground workers.

The power supply in the high-voltage room must be completely shut off, and proper safety measures must be taken before work can commence. Workshop leaders must be present on-site to ensure safety. When changing the busbar or supporting porcelain bottles, pay attention to personal and equipment safety to prevent injuries and equipment damage.

When conducting high-voltage tests on electrical equipment, a fence must be set up around the work area, with a sign hanging on the fence with "No Entry, High Voltage Danger!" Guards should be stationed there. Only one work group is allowed to perform high-voltage tests on one piece of equipment within a specific electrical connection. A grounding wire should be installed on the maintenance side of the disconnection point, and a sign with "No Entry, High Voltage Danger!" should be hung on the high-voltage test side, facing the maintenance work area.

Individuals are not allowed to enter high-voltage compartments or protective fences alone. The distance between individuals and live parts must be equal to or greater than the value specified. When there are high-voltage equipment near the work area, fences should be set up around the work area with corresponding signs hanging. There must be dedicated personnel to supervise the work, and operators must use insulating tools while standing on insulating mats. Short-circuiting of live terminals must be avoided. Measures should be taken to ensure that the secondary side of potential transformers does not short-circuit and the mutual secondary side of current transformers does not open circuit.

When the distance between workers and live parts of high-voltage equipment equals the specified safety distance, the following operations are allowed to be performed on the high-voltage equipment without power outage: replacing silicone and taking oil samples. In any circumstances, workers must maintain the specified safety distance from live parts. The safety level of workers and supervisors must not be lower than level two.

Before carrying out any work related to high-voltage electrical equipment, power workers need to thoroughly read the equipment user manual. During the work process, appropriate safety precautions should be taken, and relevant operating specifications must be strictly followed to protect one's own life. Fixed placement of instruments should be done according to the instrument's functional classification, as shown in Figure 1-3.

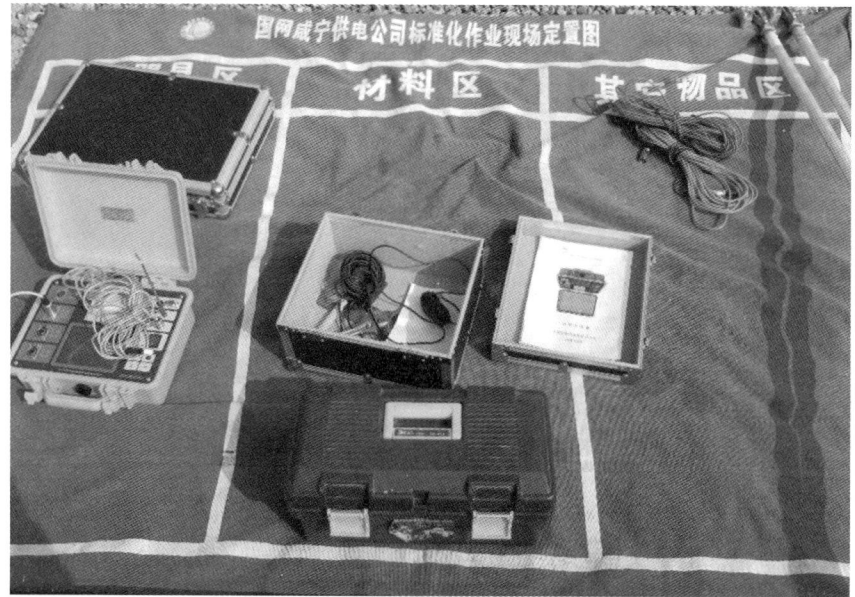

Figure 1-3 Layout diagram of standardized worksite

During high-voltage testing, the test supervisor should be an experienced staff. Before starting the test, the test supervisor should provide detailed instructions on safety precautions to all test personnel and assign dedicated personnel for supervision. The metal casing of the test equipment should be reliably grounded using multi-stranded soft bare copper wire with a cross-sectional area not less than 25 mm^2. The high-voltage leads should be kept as short as possible and supported securely with insulation materials if necessary. The power switch of the test equipment should be a double-pole switch with a clearly visible disconnect point. In the operation circuit of the test equipment, in addition to the power switch, a zero position switch should be connected in series, and an overload automatic tripping device should be installed. Before and after the withstand voltage test of large capacitor equipment or capacitors, sufficient grounding and short-circuit discharge should be carried out.

When performing voltage testing on high-voltage equipment or installing/removing grounding wires, it is necessary for two persons to work simultaneously. The operator and the supervisor must wear insulating boots and safety helmets, and the operator must wear insulating gloves. During voltage testing, an appropriate and qualified electroscope with suitable voltage ratings must be used. The electroscope should first be tested on energized equipment to confirm its proper functioning, and then voltage testing can be performed on de-energized equipment. Finally, a retest should be conducted on energized equipment. During voltage testing, all incoming and outgoing lines of the equipment being tested should be checked. High voltage testing can eliminate errors such as stop wrong electricity or no power outage, and prevent hanging ground wire in the case of live, as shown in Figure 1-4, Figure 1-5.

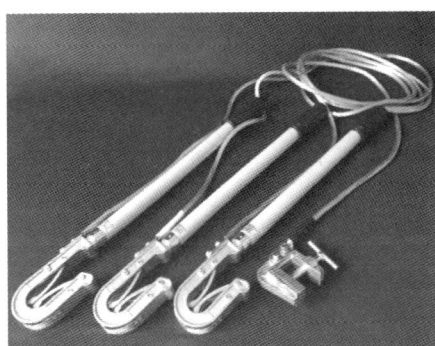
Figure 1-4 10 kV grounding wire

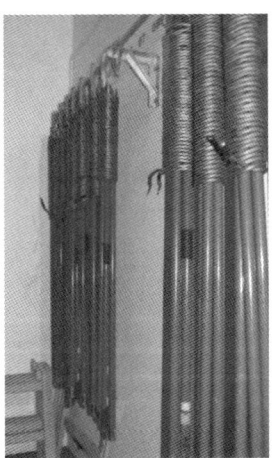
Figure 1-5 220 kV high-voltage grounding wire

There should be personnel supervision during the installation and removal of grounding wires, as shown in Figure1 1-6. The installed grounding wire should have good contact and reliable connection. The grounding wire should be connected to the grounding end first, and then to the conductor end during installation. The sequence is reversed for the removal of the grounding wire.

Figure 1-6 Grounding wires operation

II. Notes

Installation and removal of grounding wires with one person operating and another person supervising. The grounding wire shall be used with special clips, firmly connected, good contact, strictly prohibited winding. The grounding wire should use bare copper soft strand with a cross-sectional area of not less than 25 mm^2 (not less than 95 mm^2 for DC system), and there should be no broken strands, loose strands and joints.

After connecting the grounding wire for voltage testing, "Prohibited from Closing Operation, Personnel Working" signs should be hung on the operating handles of circuit breakers and disconnect switches. While working on indoor equipment, barriers should be installed, and "No Entry, High Voltage Danger!" signs should be displayed. It is prohibited for anyone to work alone in high-voltage rooms, high-voltage cabinets, or container equipment. While working in confined spaces, it is necessary to ensure good ventilation.

Module 4 Training

Exercise 1. Requirements for high-voltage equipment voltage testing:

___

Exercise 2. Requirements for installation and removal of grounding wires:

___

Worksheet 2 High Voltage Insulation Tools

Module 1 Operating Worksheet: High Voltage Insulation Tools

(Ⅰ) Insulating Tool Names	(Ⅱ) Safety Regulations
Insulating gloves, insulating boots, safety helmets, insulating mats, insulating rods, insulating pliers, electric test pens, grounding wires, barrier, signage	(1) Power-off, voltage testing, and grounding before testing. (2) Personal and environmental protection measures should be taken
(Ⅲ) Preparations before Testing	(Ⅳ) Test Procedure
(1) Check the safe service period of tools. (2) Do a good job of "two-wearing and three-carrying". (3) Record temperature and humidity	(1) Power-off, voltage testing, and grounding before testing. (2) Test wiring. (3) Record data. (4) Clean up the site
(Ⅴ) Precautions	(Ⅵ) Test Operations
(1) Set up safety barriers. (2) Take necessary precautions for working at heights	(1) Inspect insulating gloves and insulating boots. (2) Wear insulating uniform
(Ⅶ) Tool Storage	(Ⅷ) Digital resources
(1) Store tools in a dry and cool environment. (2) Keep them on dedicated equipment racks or in specialized cabinets	**Fully Automatic Insulated Tools Tester** **Safety Rope Insulation Withstand Voltage Test Device**

Module 2 Follow Me

Ⅰ. Classification of Insulation Tools

High voltage safety tools are the basic equipment used to directly protect the safety of electricians. High voltage safety tools are classified into insulation safety tools and general protective safety tools, and insulation safety tools can be further divided into basic safety tools and auxiliary safety tools.

Basic safety tools: The insulation strength should be able to withstand the working voltage for a long time and ensure the personal safety of the workers when overvoltage occurs at the working voltage level.

Auxiliary safety tools: If the insulation strength cannot withstand the working voltage of electrical equipment or circuits, it is necessary to strengthen the protection function of basic safety tools to prevent the harm of contact voltage, step voltage, and arc burning to the operators.

In high-voltage insulation safety tools, basic safety tools include insulating rods, insulating pliers, and voltage testers; auxiliary safety tools include insulating gloves, insulating boots, insulating mats, insulating platforms, and insulating blankets, etc. In low-voltage insulation safety tools, basic safety tools include insulating gloves, tools with insulating handles, and low-voltage voltage testers. Auxiliary safety tools include insulating tables, insulating mats, insulating shoes, and insulating boots, etc. Even for the same insulating gloves, the classification is different in high and low voltage applications.

Portable grounding wires, temporary barriers, signs, warning signs, protective goggles, safety belts, wooden ladders, and foot buckles, etc., are general safety tools that prevent workers from being shocked, arc burned, and falling from heights. They are not insulation tools themselves.

II. Overview of Insulation Tool Functions

The names, functions, and figures of common testing safety tools are shown in Table 1-1.

Table 1-1　Names, functions, and examples of common testing safety tools

Classification	Names and functions of insulating tools	Example
Insulating protective equipment	Safety helmets (Personal protective equipment for head protection of workers in situations where there is a risk of falling objects or personnel)	
	Insulated Boots (Basic personal safety equipment for low voltage circuit testing)	

Continued

Classification	Names and functions of insulating tools	Example
Insulating protective equipment	Insulated Gloves (Basic personal safety equipment for low voltage circuit testing)	
	Insulating Mat (Auxiliary safety equipment for personnel and equipment during high voltage testing)	
	Insulating Blanket (Auxiliary safety equipment for personnel and energized equipment during live operations)	
	High Voltage Insulating uniform (Auxiliary safety equipment for personnel and energized equipment during live operations)	Hood — Insulating gloves — Insulating jacket — Insulating pants — Insulating boots

Continued

Classification	Names and functions of insulating tools	Example
Insulating protective equipment	High Voltage Insulating uniform (Auxiliary safety equipment for personnel and energized equipment during live operations)	
High voltage power test and grounding device	High Voltage Electroscope (Used to check if electrical equipment or circuits are energized and detect the presence of high-frequency electric fields)	
	Grounding wire (Safety tool used during testing or construction to prevent electric shock hazards)	
	Insulating clamps (Used for installation and removal of high voltage fuses in 35 kV and below power system)	

Continued

Classification	Names and functions of insulating tools	Example
High voltage power test and grounding device	High Voltage DC discharge Rods (In the high voltage test, especially after the DC withstand voltage test, the accumulated charge on the sample is discharged to the ground to ensure personal safety)	
	High Voltage Insulation Operating Pole (Used for opening high voltage disconnector switches or drop-out type fuses, as well as insulation and testing measurements on high voltage equipment)	
	Phase detector (Used to detect high voltage, phase sequence verification of three-phase lines, and whether the power supply on the loop network side is in phase)	

Module 3 Workshop

Ⅰ. Insulating glove inspection

Insulating gloves can prevent a certain degree of electric shock injury. When the current leaked from the human body is within the safe range, it is safe. However, if the leakage current is too large, it can also cause human injury.

Before using insulating gloves, check whether they have exceeded their effective use period.

Conduct an external inspection before use. Use a dry cloth to wipe off the surface dirt and dust on the insulating gloves. Check the surface of the insulating gloves for scratches and inspect the insulation rubber for aging and bonding. If damage such as stickiness, cracks, leaks, bubbles, and brittleness is found, it is forbidden to use the gloves.

Inspection method: Roll the gloves toward the fingers. If the fingers bulge without leaking, they are good; alternatively, a portable insulation glove tester can be used. Inflate the gloves with the tester and check for leaks.

During the use of insulating gloves, the cuffs of the uniform must not be exposed outside the gloves. After used, clean and air dry the gloves, keep them dry and clean, and sprinkle some talcum powder to avoid adhesion. Insulating gloves should be stored in a dry and cool insulating safety tool cabinet. It should not be too cold or too hot, and the temperature should be maintained between 15 °C–35 °C with a relative humidity of 5%–80% to prevent polymer aging and reduce insulation performance. Keep the gloves separate from other tools and do not stack any objects on them to prevent puncturing the gloves.

Insulating shoes (boots) mainly undergo external size inspections. If there are damages on the surface, the shoe soles have worn out slip-resistant teeth, and the outer sole has worn through the insulation layer, they cannot be used again.

II. Requirements for wearing live operation insulating uniform

There is a difference between insulating uniform and conductive uniform. Insulating uniform can withstand high voltages below 7 kV, and it is resistant to high voltage, flame retardant, acid and alkali-resistant. Insulating uniform is made of nylon-coated fabric with insulation properties, primarily used for physical protection during live operations.

Conductive uniform, also known as live operation shielding clothing or equipotential voltage clothing, is worn to form an equipotential shielding surface on the outer surface of the human body in a high-voltage electric field. It protects the body from the dangers of high-voltage electric fields and electromagnetic waves, primarily used when entering a high-voltage electric field during the maintenance of high-voltage lines. Conductive uniform is made of uniform conductor material and fiber material. The complete set of shielding uniform should include a jacket, pants, hat, socks, gloves, shoes, as well as corresponding connecting wires and connectors.

(1) Before wearing inspect the protective equipment. Prior to use, pressurize the high- voltage insulating gloves and boots with air to check for pinhole defects. Inspect the insulating sleeves, shoulder capes, and insulating uniform for puncture holes and scratches before use. If there are significant defects in the protective equipment, its use is strictly prohibited.

(2) Operation period operator must wear insulating helmets, insulating boots, insulating uniform, insulating gloves, and outer protective gloves on the ground before entering the insulating bucket. On-site safety inspectors will conduct inspections. Refer to Figure 1-7 for illustration.

(3) When removing insulating gloves and insulating caps after the live operation, personnel should pay attention to overhead live wires and maintain a safe distance from live conductors in the vicinity.

Figure 1-7　Performing live operations while wearing insulating uniform

Module 4　Training

Exercise 1. Inspection contents of high-voltage insulating rods:

___

___

___

Exercise 2. Requirements for inspection of high-voltage electroscope:

___

___

___

Worksheet 3 High Voltage Testing and Instruments

Module 1 Operating Worksheet: High Voltage Testing Items

(Ⅰ) High voltage insulation Test Items	(Ⅱ) High voltage characteristic test device
(1) Insulation resistance testing (2) DC resistance testing (3) Dielectric loss testing (4) Fully automatic capacitance testing (5) Gas leak detection (6) Micro-water testing (7) Chromatography analysis (8) Arrester counter motion testing (9) Partial discharge measurement (10) Ground resistance measurement (11) Ground network conductivity testing (12) AC frequency withstand voltage testing (13) Series resonance testing (14) Triple frequency induced withstand voltage testing (15) Insulating oil withstand voltage testing (16) Insulating tool withstand voltage testing	(1) DC resistance testing instruments (2) Loop resistance testing instruments (3) On-load tap changer switch characteristic testing instrument for transformers (4) Transformer characteristic testing instruments (5) Instrument Transformer characteristic testing instruments (6) High-voltage switch characteristic testing instrument (7) Transformer winding deformation testing instrument
	(Ⅲ) Digital resources
	Insulation Resistance Tester **DC High Voltage Generator Tester** **Dielectric Loss Tester** **Series Connected AC Test Transformer**

Module 2 Follow Me

High-voltage equipment insulation defects are mainly classified into two types: centralized defects and distributed defects. Centralized defects refer to specific issues such as moisture ingress, mechanical damage, insulation internal bubbles, and porcelain insulator rupture, which pose significant safety risks. Distributed defects pertain to overall insulation performance degradation resulting from moisture ingress, overheating, and prolonged overload. It is a common form of insulation deterioration that progresses slowly and can be detected through periodic testing.

The purpose of high-voltage testing is to analyze and evaluate various performance aspects based on test results, eliminate and prevent latent defects, promptly identify and address equipment aging and degradation issues, and enhance equipment reliability during operation. High-voltage testing can be divided into insulation testing and characteristic testing, depending on the nature of the test items. Insulation testing involves inspections of electrical equipment insulation condition, including external insulation inspection, insulation condition data testing, and withstand voltage tests. Characteristic testing, on the other hand, refers to electrical tests conducted beyond insulation testing. Its objective is to verify whether the technical characteristics of electrical equipment meet the requirements specified in relevant technical regulations, ensuring normal operation. Transformers, instrument transformers, circuit breakers, lighting arresters, and similar electrical equipment require both insulation testing and characteristic testing. Insulators, power cables, and similar equipment typically undergo only insulation testing without characteristic testing.

Ⅰ. Classification of high-voltage tests based on failure strength

Insulation testing can be categorized into non-destructive testing and destructive testing based on the level of risk to the insulation of power equipment. Non-destructive testing is performed first and only upon passing, destructive testing can be carried out. In case of failure, insulation restoration procedures such as drying and surface cleaning are carried out.

1. Non-destructive testing

Non-destructive testing refers to the measurement of insulation characteristics of the test specimen using test voltages lower than the rated voltage of the electrical equipment to determine if there are any defects within the insulation. This method does not damage the insulation. Common non-destructive testing methods include insulation resistance and polarization index measurement, dielectric loss measurement, direct current leakage current measurement, and most characteristic tests. As shown in Figure 1-8.

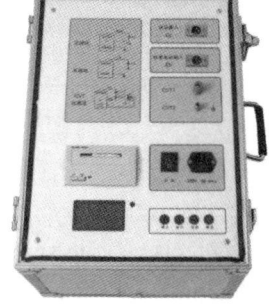

(a) Insulation resistance tester   (b) Dielectric loss tester   (c) DC high voltage generation device

Figure 1-8　Non-destructive testing device

2. Destructive testing

Destructive testing refers to the application of test voltages that are much higher than the normal operating voltage of electrical equipment for high voltage testing, to assess the ability of the electrical equipment to withstand overvoltage and insulation clearance. If the insulation clearance of the electrical equipment does not meet the requirements specified by technical standards, the insulation will break down during the high-voltage test, causing damage. Therefore, this test is called destructive testing. High voltage testing includes DC withstand voltage test, AC withstand voltage test, and impulse withstand voltage test. As shown in Figure 1-9.

(a) Power frequency AC withstand voltage instrument   (b) AC resonant withstand voltage instrument

Figure 1-9   Destructive testing device

II. Classification of High Voltage Tests by Life Cycle

High voltage tests can be divided into three categories: factory tests, commissioning tests, and preventive tests. Factory tests refer to a series of tests conducted before delivering the new equipment to the customer, to ensure that the product performance meets the delivery standards. Commissioning tests are conducted jointly by the manufacturer and the owner before the new equipment is put into operation. Preventive tests refer to the tests carried out after the owner's equipment is put into operation. The test content of these three types is similar, but they follow different technical standards.

1. Factory Test

Factory tests are conducted by power equipment manufacturers to inspect and test each product according to relevant standards and product technical conditions. As shown in Figure 1-10, 1-11. The purpose of the tests is to examine the quality of product design, manufacturing, and processing, preventing defective products from leaving the factory. A comprehensive product qualification test report will be issued after the factory test.

Figure 1-10　Factory withstand voltage test for instrument transformer

Figure 1-11　Factory withstand voltage test for transformer

2. Commissioning Test

Commissioning test or major overhaul test refers to the test conducted by the installation department and maintenance department on new equipment or overhauled equipment according to relevant standards, product technical conditions, or regulations. The commissioning test is performed before the new equipment is put into use to check for any defects or damages during transportation. As shown in Figure 1-12. After maintenance, the equipment is tested to ensure that the maintenance quality meets the requirements.

Figure 1-12   On-site commissioning test for transformer

3. Preventive Test

Preventive test refers to the test conducted by the operating department and testing department during a certain period after the equipment is put into operation. As shown in Figure 1-13. The purpose of this test is to check whether there are any insulation defects and other defects in the operating equipment. Compared to factory testing and commissioning testing, it mainly focuses on insulation testing with fewer test items.

Figure 1-13   Preventive test for transformer

Module 3 Workshop

During high voltage tests in the field, the types of high voltage tests performed on equipment of different voltage levels may vary due to the equipment voltage and application scenarios. Common electrical equipment high voltage tests and instrument configurations are shown in Table 1-2.

Table 1-2 Common electrical equipment high voltage tests and instrument configurations

ID	Test Name	Instrument Name	Test Object	Test Purpose	Test Type	Destruct Test?
1	Insulation resistance measurement	Insulation resistance tester	Transformer, instrument transformer, circuit breaker, disconnector, lighting arrester, insulator, capacitor, power cable, bushing, etc.	Measurement of insulation resistance, absorption ratio, polarization index	Insulation test	No
2	Leakage current measurement, DC withstand voltage test	DC high voltage generator	Transformer, instrument transformers, lighting arrester, bushing, capacitor, power cable, etc.	Measurement of DC leakage current of power equipment, DC characteristics of zinc oxide surge arresters, DC withstand voltage test, etc.	Insulation test	Yes
3	Dielectric loss test	Dielectric Loss tester	Transformer, instrument transformer, lighting arrester, bushing, capacitor, power cable, etc.	Measurement of equipment dielectric loss to assess insulation condition	Insulation test	No
4	AC withstand voltage test	AC withstand voltage test instrument	Transformer, generator, motor, instrument transformer, lighting arrester, switch, insulator, capacitor, power cable, etc.	Determine if the insulation strength of the equipment meets requirements; it is the most effective and direct method to assess the insulation capability of power equipment	Insulation test	Yes
5	Series resonant withstand voltage test	Series Resonance Test Equipment	Transformer, instrument transformer, generator, motor, GIS switch, lighting arrester, insulator, capacitor, power cable and other capacitive equipment	Using adjustable reactance and resonating with the capacitive test object to generate high voltage for testing the insulation capability of the equipment	Insulation test	Yes
6	Induced voltage withstand test	Triple-frequency induction withstand voltage instrument	Transformer, electromagnetic instrument transformer	Measurement of the longitudinal insulation problem in transformers	Insulation test	Yes

Continued

ID	Test Name	Instrument Name	Test Object	Test Purpose	Test Type	Destruct Test?
7	Insulating Tool Inspection Test	Insulating tool withstand voltage test instrument	Insulating glove, insulating boot, insulating mat, insulating Rod	Check the qualification of auxiliary safety tools	Insulation test	Yes
8	Insulating oil withstand Voltage test	Insulating oil withstand voltage test instrument	Insulating oil	Detect moisture, contamination, etc. in transformer oil	Insulation test	Yes
9	Cable path tracing	Cable pace tracer	Power cable	Detect cable path, conduct pipeline survey, and perform depth measurement	Characteristic test	No
10	Cable fault location	Cable fault tester	Power cable	Using low-voltage pulse and high-voltage flashover methods to detect open circuit and high/low impedance short circuit faults in power cables	Characteristic test	Yes
11	DC resistance measurement	DC resistance tester	Transformer, instrument transformer	Measurement of changes in DC resistance value of transformers to determine if there are broken or loose windings	Characteristic test	No
12	Loop Resistance Measurement	Loop Resistance Tester	Circuit breaker, disconnect switch	Measurement of changes in loop resistance of switches to determine if there are discharge or burnout issues	Characteristic test	No
13	Characteristics of transformer on-load tap-changer Characteristics test of transformer on-load tap-switch	Transformer on-load tap switch characteristic instrument	Transformer	Measurement of transition resistance and transition time of on-load tap switch in transformers	characteristic test	No
14	Transformer characteristic test	Transformer characteristic tester	Transformer	Measurement of transformer parameters such as turns ratio, polarity, connection group, capacity, impedance, etc.	Characteristic test	No

Continued

ID	Test Name	Instrument Name	Test Object	Test Purpose	Test Type	Destruct Test?
15	Instrument transformer characteristic test	Instrument transformer characteristic tester	Potential transformer	Measurement of parameters such as turns ratio, polarity, and volt-ampere characteristics of potential transformers	characteristic test	No
16	High voltage switch characteristic measurement	High voltage switch characteristic tester	Circuit breaker, disconnect switch	Measurement of switch opening and closing time, speed, action voltage measurement, synchronization, etc.	characteristic test	No
17	Capacitance measurement	Fully automatic capacitance tester	Power capacitor	Check the capacity of power capacitors	characteristic test	No
18	Gas leak detection test	Gas leak detector	GIS	Detection of SF_6 and other gas leaks	characteristic test	No
19	Micro-Moisture Test	Micro-Moisture meter	GIS	Detection of moisture content in SF_6 and other gases	characteristic test	No
20	Winding deformation measurement	Transformer winding deformation tester	Transformer	Detection of deformation due to mechanical shock during operation of transformers	Characteristic test	No
21	Insulating oil chromatography analysis	Chromatography analyzer	Insulating oil	Detection of overheating or discharge faults inside transformers	Insulation test	No
22	Partial discharge	Partial discharge tester	Transformer, instrument transformer, circuit breaker, disconnect switch, insulator, capacitor, power cable, bushing, etc.	Measurement of the existence of partial discharges in electrical equipment	Insulation test	No
23	Ground resistance measurement	Ground resistance test	Grounding grid, grounding wire	Measurement of ground grid resistance, grounding impedance, step voltage, contact potential, etc.	Insulation test	No

Continued

ID	Test Name	Instrument Name	Test Object	Test Purpose	Test Type	Destruct Test?
24	Grounding grid measurements	Ground continuity tester	Grounder, lighting arrester	Checking the connection between the grounding lead of electrical equipment and the power grid, as well as the grounding resistance of surge arresters	Insulation test	No
25	Arrester discharge counter verification	Arrester discharge counter test instrument	Various types of arrester discharge counter	Testing the operation of various types of arrester discharge counters and calibration of ammeters	Characteristic test	No
26	Zinc Oxide lightning arrester test	Zinc oxide lightning arrester tester	Zinc oxide lighting arrester	Specialized instrument for measuring the full current, resistive current, their harmonic components, fundamental frequency reference voltage and its harmonic components, active power, and phase difference of zinc oxide lighting arresters	Characteristic test	No
27	High voltage wireless nuclear phase	High voltage wireless nuclear phase instrument	High voltage line phase	Line nuclear phase	Characteristic test	No
28	Vacuum degree measurement	Vacuum degree tester	Various types of vacuum switch	Vacuum degree measurement for various types of vacuum switches	Characteristic test	No
29	Infrared thermal imaging	Infrared thermal imaging camera	Circuit breaker, disconnector, GIS, transformer, bushing, etc.	Detection of thermal faults in electrical equipment	Insulation test	No

Tests shall be conducted on different types of power equipment, as well as power equipment of the same type but different voltage levels, in accordance with the "Preventive Test Regulations for Power Equipment" (DL/T 596–2021) as specified.

Module 4　Training

Exercise 1. Insulation tests are categorized by the degree of danger to the insulation of electrical equipment: _____

The difference between these two experiments is: _____

_____

The two test types are: _____

_____

Exercise 2. High voltage tests are categorized by the different tasks of the experiment:

_____

The three types of tests differ in the following: _____

_____

Module 5　Evaluation

Practical examination, describe the key points of the test, etc. As shown in Table 1-3.

Table 1-3　Evaluation form

Project Name	Power testing and grounding	Evaluation
Testing Instruments	Detector, grounding wire	
Test content	① Test the electricity. ② Install the ground wire. ③ Remove the ground wire	
insulating tools	Detector, grounding wire	
Potential Risks	Risk of electrocution	
Project Requirements	① Power inspection and grounding personnel should be familiar with grounding locations such as railroad skylight sections, blackout arms, etc. ② Test the electricity, disassemble and install the grounding wire. ③ During electrical verification, appropriate and qualified voltage testers must be used	
Material Preparation	Insulating gloves, insulating shoes, helmets	
Security Risk Control	① When conducting maintenance on railway power supply sectioning, phasing, isolation switches, and surge arresters, reliable connections must be made using short-circuiting wires. ② Prior to disassembly, installation, or contact with live cable heads, capacitors, and other charged equipment, an electrical verification and grounding must be conducted, and electrical verification should also be confirmed after installation and grid connection. ③ The personnel responsible for electrical verification and grounding should be familiar with the locations of roof access areas, de-energized arms, and grounding wires, and their supervisors should recite them during planning meetings	
Test process	Power outage → power check → discharge → hook up the ground wire	
Organizing the site	After the test is completed, restore the test site to its original state	

Project 2

High Voltage Insulation Test

Worksheet 1 Insulation Resistance Test

Module 1 Operating Worksheet: Insulation Resistance test

(Ⅰ) Test Name and Instrument	(Ⅱ) Test Objects
Insulation resistance test **Insulation Resistance Tester**	Transformer, instrument transformer, high-voltage switch, lightning arrester, power capacitor, reactor, power cable, insulator, bushing, GIS, generator, and motor
(Ⅲ) Test Purpose	(Ⅳ) Measurement Steps
(1) Effectively detect local or overall moisture in insulation, moisture or dirt on component surfaces, and penetrative defects. (2) Effectively detect defects such as insulation penetration short circuit, porcelain bottle damage, connecting wire connection to the casing, and copper wire bridging. (3) Effectively detect grounding defects in the manufacturing of equipment, such as iron core and bracket grounding defects	(1) Disconnect the power supply of the tested product, discharge the tested product, and remove the original connections. (2) Clean the dirt from the outer insulation surface of the tested product. (3) Conduct open circuit and short circuit tests with a megohmmeter. (4) The "E" terminal of the megohmmeter is connected to the grounding end of the tested product, and the "L" terminal is connected to the high voltage end. (5) After each test, discharge the tested product

Continued

(Ⅴ) Precautions	(Ⅵ) Technical Standards
(1) During measurement, all leads of the non-tested windings should be shorted and grounded to avoid measurement errors caused by residual charges in each winding. (2) Insulation resistance needs to be temperature-calibrated. The absorption ratio and polarization index do not require temperature calibration. (3) After temperature calibration, compare the test results with the factory test values	(1) The absorption ratio should not be lower than 1.3, and the polarization index should not be lower than 1.5. (2) For transformers with an absorption ratio less than 1.3, where it is difficult to draw a conclusion, the measurement of the polarization index can be supplemented as a basis for comprehensive judgment. (3) The measured value of the polarization index should not be lower than 1.5
(Ⅶ) Result Judgment	(Ⅷ) Digital resources
(1) During installation, the insulation resistance value R_{60s} should not be lower than 70% of the insulation resistance measurement value obtained during factory testing. (2) During preventive tests, the insulation resistance value R_{60s} should not be lower than 50% of the measurement value obtained before installation or major repairs prior to commissioning. For a 500 kV transformer, at the same temperature, its insulation resistance should not be less than 70% of the factory value, with a minimum resistance value of 2,000 MΩ at 20 °C. (3) The "Regulations" stipulate the use of absorption ratio and polarization index to assess the insulation condition of large transformers. The measured value of the polarization index should not be lower than 1.5. (4) The absorption ratio is related to temperature. With good insulation, as the temperature increases, the absorption ratio increases. However, in the case of poor oil or paper insulation, as the temperature increases, the absorption ratio decreases	**GIS Insulation Resistance Test** **Power Cable Insulation Resistance Testing**

Module 2 Follow Me

Ⅰ. Overview of insulation resistance

Measuring the insulation resistance, particularly the absorption ratio and polarization index of the transformer, is highly sensitive in assessing the overall insulation condition. This method effectively identifies issues like moisture, surface contamination, and through-hole defects in the transformer insulation. It can also detect problems such as short circuits, grounding, or cracks in the porcelain insulation. Insulating materials within the transformer, including the internal iron core, clamps, and through-bolts, play a crucial role in ensuring insulation. The insulation in the winding sections can withstand high voltage levels. When the insulation of the transformer is compromised due to piercing, damage to the ceramic insulation, connection of the lead wire to the casing, or the occurrence of metal short-circuit faults, there will be a noticeable change in the insulation resistance. Moreover, the change in insulation resistance before and after equipment drying is significantly larger than the change in the dissipation factor of the dielectric. For instance, in the 7,500 kVA transformer, the dissipation factor may change by 2.5 times before and after drying, while the insulation resistance may change by more than 40 times. This change is quite remarkable. Therefore, evaluating the insulation resistance of components like the iron core can effectively detect overall moisture in the transformer insulation, surface contamination of components, and concentrated through-hole defects.

The absorption ratio K refers to the ratio of the insulation resistance value at 60 seconds to the insulation resistance value at 15 seconds. The polarization index PI refers to the ratio of the insulation resistance value at 10 minutes to the insulation resistance value at 1 minute during testing.

$$K=R_{60}/R_{15} \tag{2-1}$$

$$PI=R_{600}/R_{60} \tag{2-2}$$

The polarization index is an important parameter for assessing whether large power equipment is affected by moisture. For the large transformer, generator, power cable, and parallel capacitor, due to their large sample capacitance and composite insulation materials, the absorption current decays slowly, and the polarization process often cannot be completed within 1 minute. Therefore, it is necessary to measure for 10 minutes, and compared to factory tests, the insulation condition is judged using the polarization index (PI). Take transformers as an example, due to the improvement in drying processes, the insulation resistance of transformers is increasing and can generally reach several tens of thousands of megaohms, causing the polarization process of the transformer to become longer. Therefore, the polarization index should be measured, and the absorption ratio test results should not be used to determine if the transformer is qualified. When the insulation resistance of the transformer exceeds 10,000 MΩ, there is no need to consider the absorption ratio or polarization index. Therefore, when conducting insulation resistance, absorption ratio, or polarization index testing, the results should be analyzed and judged in combination with specific electrical equipment.

In the "Regulations", it is stipulated that the absorption ratio and polarization index are not converted by temperature.

II. Test Standards

According to the "Regulations on Preventive Testing of Power Equipment", it is stipulated that the absorption ratio should not be less than 1.3 or the polarization index should not be less than 1.5 as the compliance standard. As shown in Table 2-1, 2-2. After measuring insulation resistance, it should be uniformly converted to 20 °C and compared with the factory value or the value from the previous measurement cycle.

Table 2-1 Reference standards for assessing insulation condition based on the polarization index.

Status	Polarization Index (*PI*)
Hazardous	Less than 1.0
Poor	1.0–1.1
suspicious	1.1–1.25
fair	1.25–2.0
good	Greater than 2.0

Table 2-2 Qualified standard for insulation resistance test

ID	Power equipment	Qualified standard	Remarks
1	Transformer	(1) Converted to the same temperature and without significant changes from the previous measurement. If there is no initial value, the value should be 800 MΩ. (2) The absorption ratio (within the range of 10–30 °C) should be not less than 1.3, or the polarization index should be not less than 1.5	For the primary winding, use a 2.5 kV or 5 kV megohmmeter; for the secondary winding, use a 1 kV megohmmeter
2	Instrument transformer	The insulation resistance of the winding should not have significant changes compared to the initial value and previous data. The insulation resistance of the tap grouding end of capacitive current transformers to ground should generally not be lower than 1,000 MΩ, and the insulation resistance of the winding of the potential transformer should be not less than 3,000 MΩ (35 kV), 5,000 MΩ (110 kV), and 10 MΩ for secondary insulation resistance	For the primary winding, use a 2.5 kV or 5 kV megohmmeter; for the secondary winding, use a 1 kV megohmmeter
3	Circuit breaker, disconnect switch	(1) For the vacuum circuit breaker, air circuit breaker, and SF_6 circuit breaker, the insulation resistance of primary circuit components such as porcelain sleeves and pull rods to ground should be greater than 5,000 MΩ. (2) The insulation resistance value of the breaking and closing coils of the vacuum circuit breaker and the closing contactor coil should not be less than 10 MΩ. (3) Other reference standards should follow the manufacturer's technical specifications	For the primary loop measurement, use a 2.5 kV megohmmeter; for the secondary loop measurement, use a 1 kV megohmmeter
4	Lightning arrester	Metal Oxide Surge Arrester: (1) Above 35 kV, not less than 2,500 MΩ. (2) 35 kV and below, not less than 1,000 MΩ	Use a 2.5 kV megohmmeter for measurement
5	Power cable	(1) Above 35 kV, not less than 500 MΩ. (2) 35 kV and below, not less than 300 MΩ. (3) Other reference standards should follow the manufacturer's technical specifications	(1) For the cable with a rated voltage of 0.6/1 kV, a 1 kV megohmmeter is used. (2) For the cable above 0.6/1 kV, a 2.5 kV megohmmeter is used. (3) For the cable with a voltage rating of 6 kV and above, a 5 kV megohmmeter can also be used

Continued

ID	Power equipment	Qualified standard	Remarks
6	Bushing	The insulation resistance value of the main insulation should not be lower than 10,000 MΩ. The tap grounding end insulation resistance value of the capacitive-type bushing should not be less than 1,000 MΩ	Use a 2.5 kV megohmmeter for measurement
7	Reactor	Generally, not less than 1,000 MΩ	Use a 2.5 kV megohmmeter for measurement
8	Capacitors	Generally, not less than 2,000 MΩ	Use a 1 kV megohmmeter for the series-connected capacitor, and a 2.5 kV megohmmeter for others. The single bushing capacitor is not measured
9	Insulator	The insulation resistance of each component of the pin-type insulator and each disc of suspension insulators should not be less than 300 MΩ. For the 500 kV suspension insulator, the insulation resistance should not be lower than 500 MΩ	Use a megohmmeter with a voltage of 2.5 kV or higher for measurement
10	Generator	(1) The difference in insulation resistance values between phases or branches should not exceed 100% of the minimum value. (2) Absorption ratio or polarization index: The absorption ratio of asphalt-impregnated and tape mica insulation should not be less than 1.3, or the polarization index should not be less than 1.5. The absorption ratio of epoxy powder mica insulation should not be less than 1.6, or the polarization index should not be less than 2.0. The water-immersed stator winding may have its own specified requirements	

Module 3 Workshop

Ⅰ. Measurement methods

During the measurement, record the ambient temperature and humidity. Measure the insulation resistance and absorption ratio sequentially for each winding with respect to ground and between other windings. Short-circuit all leads of the winding being tested, and short-circuit and ground all leads of the non-tested windings. This allows for measuring the insulation condition between the tested winding and ground, as well as between the tested winding and the non-tested windings, while avoiding the influence of residual charges in the non-tested windings on the measurement.

Figure 2-1 shows the functional diagram of the insulation resistance tester, Figure 2-2 shows the wiring diagram for measuring the insulation resistance of the high-voltage winding of the high-voltage generator.

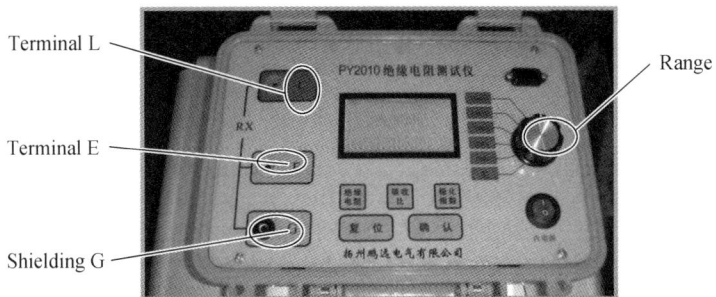

Figure 2-1　Insulation resistance tester function diagram

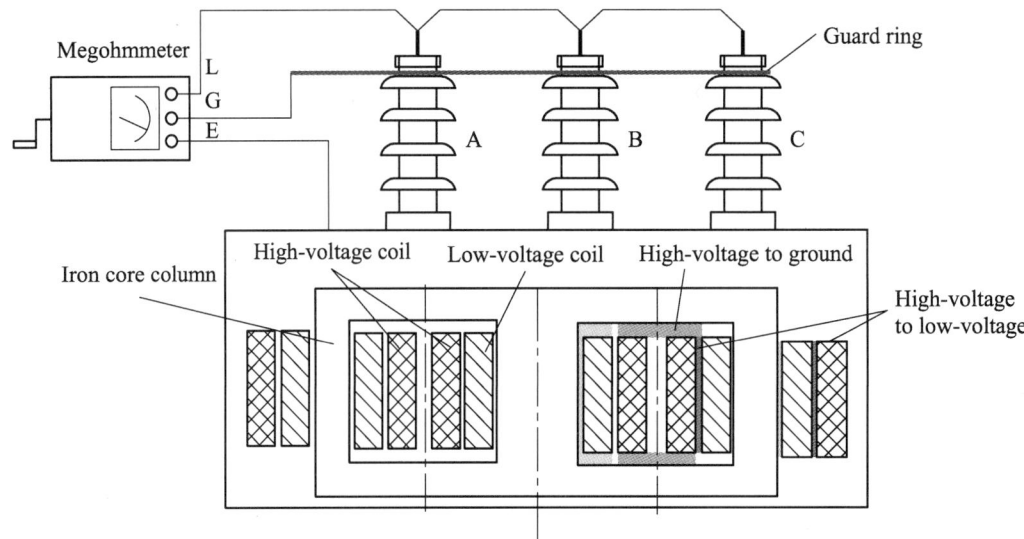

Figure 2-2　Wiring diagram for insulation resistance measurement of three-Phase transformer high-voltage winding

When measuring the insulation resistance of windings rated at 1 kV and below, use a megohmmeter with a range not exceeding 0.5 kV. For windings with voltages of 2.5 kV and above, a megohmmeter with a voltage of 1 kV or 2.5 kV can be used. For windings rated at 10 kV and above, use a 5 kV insulation resistance tester and record the top oil temperature.

Analysis of insulation resistance measurement results is conducted using a comparative method, primarily relying on comparing the results with previous test results of this transformer. Generally, the switching test value should not be lower than 70% of the factory test value. When converting insulation resistance to the insulation resistance at 20 °C, transformers up to 220 kV should have a value not less than 800 MΩ, and the transformer of 500 kV should have a value not less than 2,000 MΩ. The absorption ratio should not be lower than 1.3.

When measuring the absorption ratio K of the transformer, the iron core must be grounded. If the iron core is not grounded, insulation materials such as cardboard between the transformer windings and the casing are introduced in series, causing an increase in the measured insulation value R_{15}, resulting in a decrease in the absorption ratio, which can easily lead to an unqualified result.

II. Technical standards

For the large power transformer of 35 kV and below, the absorption ratio should not be lower than 1.3. For the large power transformers with voltages equal to or higher than 60 kV, the absorption ratio should be controlled not lower than 1.5. In the commissioning test of the power industry, the corresponding requirement for the absorption ratio is not lower than 1.2 and 1.3, respectively.

III. Judgment and Analysis

Insulation resistance can partially reflect the insulation condition of windings. However, it is greatly influenced by factors such as insulation structure, operation mode, environment, equipment temperature, insulation oil quality, and measurement errors. Therefore, during installation, the insulation resistance value R_{60s} should not be lower than 70% of the measured insulation resistance value during the factory test. During preventive tests, the insulation resistance value R_{60s} should not be lower than 50% of the measured value after installation, major repairs, and prior to being put into operation. For the 500 kV transformer, under the same temperature conditions, its insulation resistance should not be lower than 70% of the factory value, and the minimum resistance value at 20 °C should not be less than 2,000 MΩ.

IV. Special Notices

Due to the designed shielding loop in the megohmmeter's internal circuitry to ensure measurement accuracy, it is important to correctly connect the circuit. The L-terminal of the line should be connected to the conductive part of the test specimen insulated from the ground, while the E-terminal should be connected to the grounding end of the test specimen. Interchanging these connections is not allowed as it may result in measurement errors.

Selecting a megohmmeter with a measuring voltage that matches the rated voltage of the specimen is crucial. The breakdown voltage of the insulation material is related to the applied voltage, and the measurement accuracy of insulation characteristics also depends on the applied voltage. If the measuring voltage is too low, it will not accurately measure the insulation condition, and if it is too high, it may damage the insulation of the specimen.

Temperature and humidity should be recorded during the measurement. The absorption ratio is temperature-dependent. For very good insulation, the absorption ratio increases with temperature. For poor insulation such as oil or paper insulation, the absorption ratio decreases with temperature. In equipment with severe moisture, the insulation resistance varies significantly with temperature. If the equipment transitions from operation to shut down, it should be measured after the internal insulation has cooled down completely, and the insulation resistance should be converted to the insulation resistance at 20 °C.

For new or Repaired oil-immersed transformers, It will take dozens of hours to wait for the oil circulation and bubble escape before measuring the insulation resistance. For the transformer with a capacity of 8,000 kVA or above, they should be left standing for at least 20 hours, while other lower-capacity transformers require a standing time of at least 5 hours. The quality of oil in the transformer directly affects the insulation resistance value. The better the oil quality, the higher the insulation resistance and the higher the absorption ratio.

For the capacitive bushing and current transformer, if they are exposed to moisture and the moisture content is higher than the specific gravity of transformer oil, water will accumulate at the bottom of the tap shielding. In this case, it is necessary to disconnect the grounding of the tap shielding and measure the insulation resistance between the tap shielding and the ground. If the insulation resistance is significantly reduced, it can be determined whether the capacitive bushing and current transformer have been affected by moisture. Refer to Figure 2-3 for details.

Figure 2-3   110 kV capacitive current transformer

The insulation resistance of the power cable is greatly affected by temperature and should be converted based on the soil temperature, not the ambient temperature. If the insulation resistance of the cable is poor, it can generally be measured through a leakage current test. By measuring the insulation values between cable phases to ground or between phases, it is possible to determine if the cable has any damage and classify it as either a high-resistance fault or a low-resistance fault.

Module 4   Training

Why measure insulation resistance?

Worksheet 2 Leakage Current and DC Withstand Voltage Test

Module 1 Operating Worksheet: Leakage Current and DC Withstand Voltage

(Ⅰ) Test Items and Instruments	(Ⅱ) Test Objects
DC leakage current and withstand voltage test **DC High Voltage Generator**	Power transformer, instrument transformer, zinc oxide lighting arrester, cable, GIS, switch, busbar, bushing, generator, etc.
(Ⅲ) Test Purposes	(Ⅳ) Measurement steps
(1) Effectively detect concentrated defects such as material selection, welding, loose connections, broken wires, cracks, and damages in power equipment. (2) Detect permeability defects or insulation degradation caused by overall moisture in power equipment	(1) Connect the test leads as required and take necessary safety measures. (2) Power on, press the high-voltage enable button, and apply high-voltage measurement. (3) Record the steady-state value of the DC leakage current. (4) Reduce the voltage and disconnect the power. (5) Discharge
(Ⅴ) Precautions	(Ⅵ) Technical Standards
(1) Ensure reliable wiring and grounding, and use an appropriate range for the microammeter. (2) Minimize the impact of surface dirt on the leakage current of the test object. (3) Connect voltage and current separately with short leads. (4) Avoid high voltages due to sudden current interruptions. (5) Fully discharge after completion of the test	(1) The difference in leakage current should not exceed 100% of the minimum value, or the three-phase leakage current should be below 50 µA. (2) There should be no significant change compared to previous test results
(Ⅶ) Result Judgments	(Ⅷ) Digital Resources
(1) In general, the measured value for the current year should not exceed 150% of the previous year's measurement. If the data gradually increases year by year, pay attention to the gradual insulation deterioration. However, if the value suddenly increases compared to previous years, there may be serious defects that need to be investigated. (2) When the leakage current is too large, the test object, test leads, shielding, and voltage should be checked and external factors should be ruled out before drawing conclusions about the test object. When the leakage current is too small, it may be caused by problems with the wiring, insufficient voltage, or the microammeter's shunt	**Leakage Current and DC Withstand Voltage**

Module 2 Follow Me

Ⅰ. Measurement Purpose

The principle of measuring leakage current is similar to the measuring insulation resistance, and the detectable defects are roughly the same. However, since the measurement of leakage current requires a higher test voltage than insulation resistance testing, the microammeter can monitor the leakage current in the circuit with higher sensitivity. Therefore, it can detect insulation defects that are only exposed under higher voltage. The insulation resistance value calculated from the leakage current should be similar to the value measured by a megohmmeter. For example, if the leakage current of a transformer increases from 15 μA to 490 μA, a 30-fold increase, it was found that the cause was water ingress due to imperfect casing sealing.

DC withstand voltage testing can effectively detect overall defects such as insulation moisture and dirt, and can also identify local defects in insulation through the relationship curve between voltage and leakage current. DC withstand voltage testing causes less damage to insulation, requires smaller testing equipment capacity, and is convenient to carry.

AC withstand voltage testing can effectively detect more dangerous concentrated defects. It is the most direct method for determining the insulation strength of electrical equipment and has a decisive significance in determining whether electrical equipment can be put into operation. It is also an important means to ensure equipment insulation level and prevent insulation accidents. Both DC withstand voltage testing and AC withstand voltage testing can effectively detect insulation defects, but they have their own characteristics. Therefore, these two methods cannot replace each other. If necessary, both methods should be carried out simultaneously to complement each other.

Ⅱ. Test Data Standards

Taking the example of the zinc oxide arrester, the DC leakage current test is conducted. The required test data for U_{1mA} and $0.75U_{1mA}$ of the zinc oxide arrester are as follows: The measured value of U_{1mA} should have a variation of less than 5% compared to the factory or initial value, and the leakage current at $0.75U_{1mA}$ should not exceed 50 μA. Otherwise, the zinc oxide arrester may be aged or dampened.

The purpose of measuring the voltage at U_{1mA} in DC is to find the critical value at which the zinc oxide arrester breaks down, check if the valve disks are damp or aged, and determine if their operational performance meets the requirements. The purpose of measuring the DC leakage current at the breakdown voltage of $0.75U_{1mA}$ is to inspect the insulation condition of the zinc oxide arrester when it has not yet broken down. The $0.75U_{1mA}$ DC voltage is generally slightly higher than the maximum working phase voltage. At this voltage, the primary objective is to test whether the long-term allowable operating current complies with regulations. These two tests are beneficial for examining the DC reference voltage of the metal-oxide arrester and its charging state during normal operation. They play a crucial role in determining the number of valve disks, assessing the rationality of the rated voltage selection, and evaluating the aging status. Because this current is directly related to the lifespan of the metal-oxide arrester, the leakage current and lifespan are inversely proportional at the same temperature.

In the field, the arrester monitor is used in series with the arrester to record the working conditions of the arrester. This is especially suitable for arrester applications in power stations and lines of 35 kV and above. It can also be used in conjunction with three-phase combined overvoltage protective devices. When the arrester operates, the discharge count is accumulated and recorded by a counter. The counter uses a three-digit electromagnetic counter that automatically resets after reaching the full count. It operates in a cyclic counting mode without clearing to zero, as shown in Figure 2-4. Leakage current test standards as shown in table 2-3.

(a) Hy5ws-17/50 zinc oxide lightning arrester
(Metal Oxide Arrester, MOA)

(b) 110 kV line arrester

(c) JCQ8 arrester monitor

Figure 2-4　Arrester

Table 2-3 Leakage current test standards

ID	Power Equipment	Qualification Standards
1	Transformer	The difference in leakage current between each phase should not be greater than 100% of the minimum value, or the leakage current of the three phases should be below 50 μA, and there should be no significant changes compared with previous test results
2	Circuit breaker, disconnect switch	The leakage current is generally not greater than 10 μA
3	Lightning arrester	The measured value of U_{1mA} should be compared with the initial value or the value specified by the manufacturer, and the change should not exceed ±5%. The leakage current under $0.75U_{1mA}$ should not exceed 50 μA
4	Power cable	For cables with a voltage of 6 kV and below, the leakage current should be less than 10 μA, and for cables with a voltage of 10 kV, the leakage current should be less than 20 μA
5	Insulator	(1) The insulation resistance of the newly installed insulator should be equal to or greater than 500 MΩ, while the insulation resistance of the insulator during operation should be equal to or greater than 300 MΩ. (2) The insulator with an insulation resistance between 240 MΩ and 300 MΩ can be classified as a low-value insulator, while those with an insulation resistance below 240 MΩ can be classified as the zero-value insulator

Module 3 Workshop

Ⅰ. DC High Voltage Generator

The sequence and location for measuring leakage current on the dual-winding and three-winding transformer are shown in Table 2-4. Figure 2-5 illustrates the connection diagram for the leakage current test principle. Figure 2-6 depicts the functional schematic diagram of the DC generator.

Table 2-4 The sequence and location for measuring leakage current in Transformer Windings

Orders	Dual-winding Transformer		Three-winding Transformer	
	Windings with applied voltage	Grounding section	Windings with applied voltage	Grounding section
1	High voltage	Low voltage, casing	High voltage	Medium, low voltage, casing
2	Low voltage	High voltage, casing	Medium voltage	High, low voltage, casing
3	Low voltage	High voltage, medium voltage, casing		

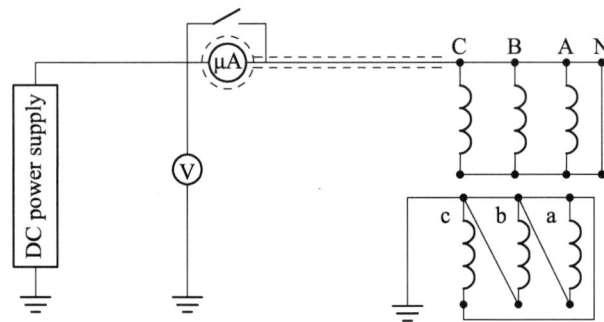

Figure 2-5 Connection principle schematic diagram for leakage current test

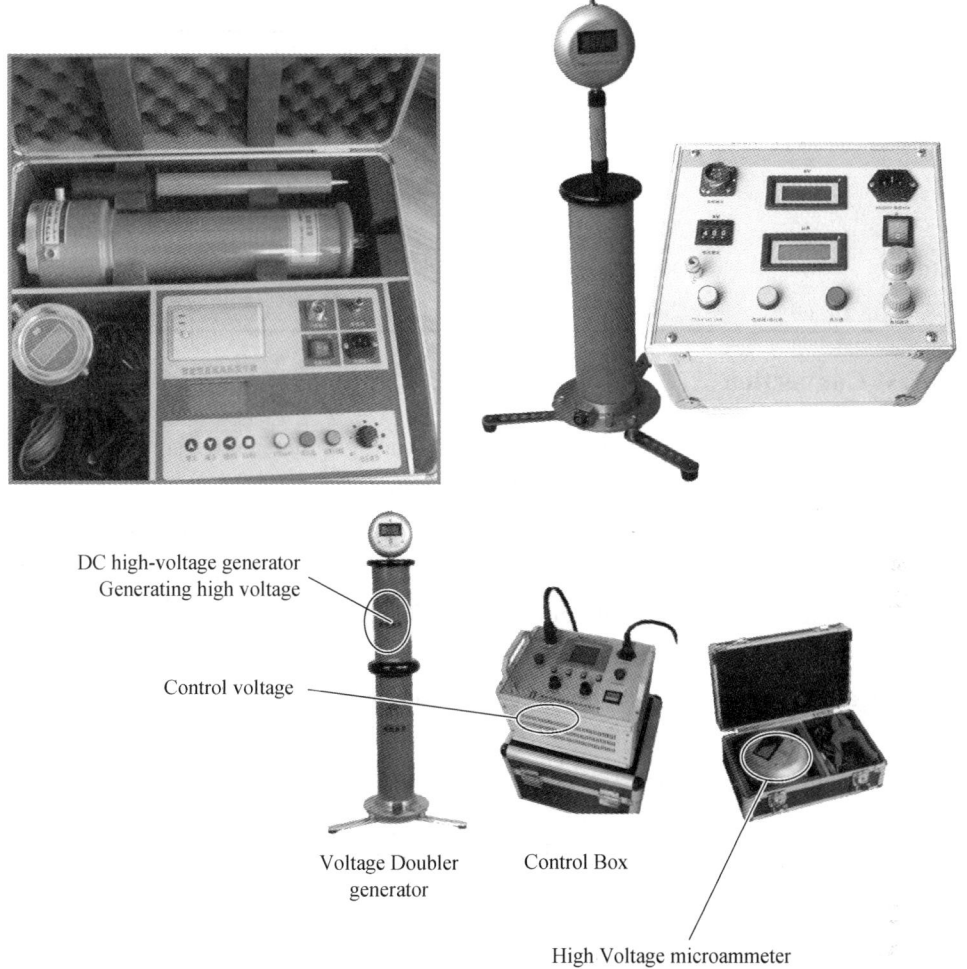

DC high-voltage generator
Generating high voltage

Control voltage

Voltage Doubler generator

Control Box

High Voltage microammeter
Accurate measurement of leakage current

Figure 2-6　Function diagram of DC generator components

The standard reference for the test voltage is shown in Table 2-5.

Table 2-5　Standard test voltages for leakage current testing

Rated Voltage of Winding /kV	3	6–15	20–35	Over 35
DC Test Voltage/kV	5	10	20	40

During the measurement, after the voltage is raised to the test voltage, the current value obtained by reading after 1 minute is the leakage current value. To ensure accurate readings, the microammeter should be connected to the high-voltage side. It should also be noted that for the transformer without oil, the voltage applied to the transformer when measuring the leakage current should be 50% of the value shown in the table. Table 2-6 provides reference values for DC leakage current in the oil-immersed power transformer winding.

Table 2-6 Reference values for DC leakage current in oil-immersed power transformer windings

Rated voltage/kV	Test voltage peak/kV	Leakage current value of the winding (μA) at the following temperatures							
		10 °C	20 °C	30 °C	40 °C	50 °C	60 °C	70 °C	80 °C
2–3	5	11	17	25	39	55	83	125	178
6–15	10	22	33	50	77	112	166	250	356
20–35	20	33	50	74	111	167	250	400	570
63–330	40	33	50	74	111	167	250	400	570
500	60	20	30	45	67	100	150	235	330

II. Test Connection

There are two methods for connecting the DC leakage current measurement: low-voltage connection method and high-voltage connection method.

The low-voltage connection method involves connecting the microammeter to the tail end of the high-voltage winding of the test transformer. Since the microammeter is on the low-voltage side, it is safer and more convenient for reading. However, this method cannot eliminate the measurement errors caused by leakage currents on insulation surfaces and corona currents on high-voltage leads. Therefore, the high-voltage connection method is commonly used for on-site tests. Figure 2-7 shows a schematic diagram of the low-voltage connection method for measuring leakage current with a microammeter.

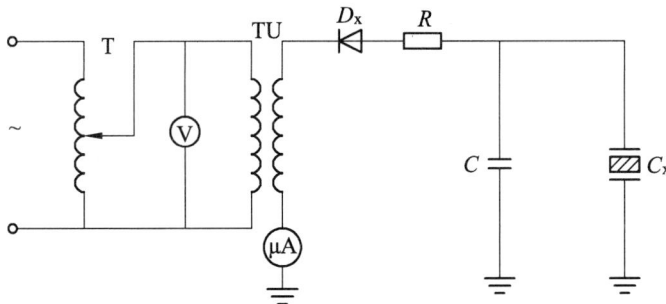

Figure 2-7 Schematic diagram of low-voltage connection method for leakage current measurement with a microameter

The high-voltage connection method involves connecting the microammeter before the test sample. In this connection method, since the microammeter is located on the high-voltage side, it is placed on the shielding frame and connected to the shielding ring of the test sample via a shielding wire. This avoids measurement errors caused by the wiring. However, placing the microammeter on the high-voltage side can be inconvenient for reading the measurements. Figure 2-8 shows the schematic diagram of the high-voltage connection method for measuring leakage current with a microammeter.

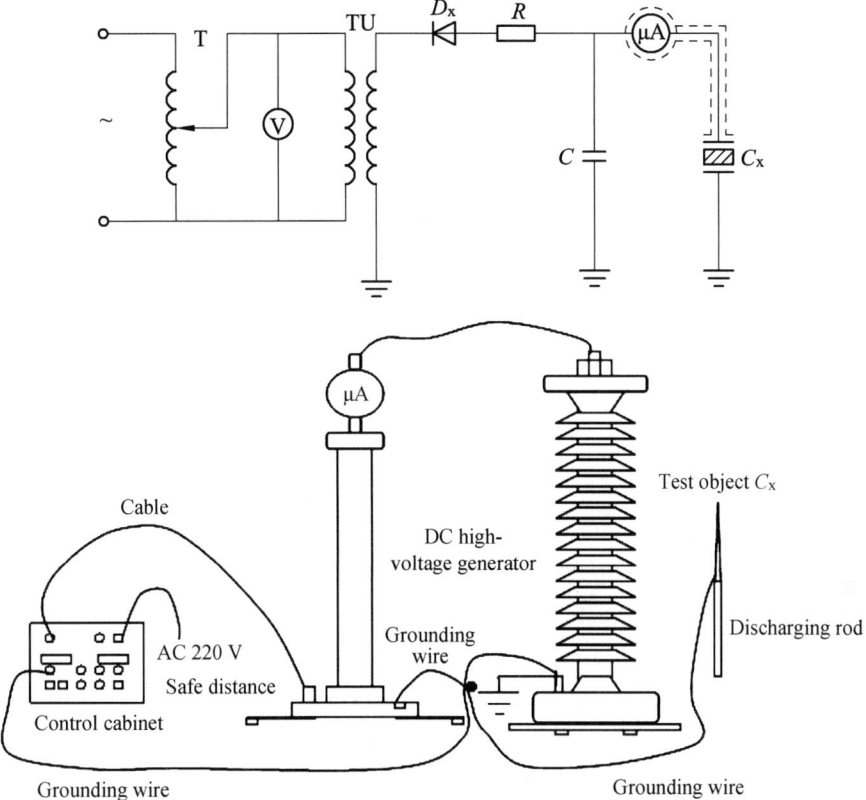

Figure 2-8 Schematic diagram of high-voltage connection method for leakage current measurement with a microammeter

III. DC Leakage Current Test Process

Test Steps:

(1) Remove or disconnect all external connections of the MOA (Metal Oxide Arrester) and ground discharge it.

(2) Clean the surface and make the connections. After verifying the correctness, remove the grounding wire and start the test.

(3) Confirm that the voltage output is at zero. Turn on the power supply and start the DC generator. Gradually adjust the voltage boost knob until the ammeter reading reaches 1 mA. Record the DC high-voltage value $U_{1\,mA}$ (the current reading from A1 meter for measuring the upper section and lower section, and from A2 meter for measuring the middle section).

(4) The system automatically reduces the voltage to $0.75 U_{1\,mA}$. Read and record the leakage current value at this point, then reduce the voltage to zero.

(5) Disconnect the test power supply and use a discharge rod to ground the test sample.

(6) Remove the test connections and clean the site.

For the test results, they are mainly judged by comparing them with previous test data. It is required that there should be no significant changes when compared with previous data. In general, the measured value of the current year should not exceed 150% of the measured value from the previous year. If the values increase year by year, Pay attention that it may indicate insulation degradation over time. If the value suddenly increases compared to previous years, it may indicate a serious defect, and the cause should be investigated.

IV. Special Notes

For on-site withstand voltage tests of the cross-linked polyethylene (XLPE) power cable, direct current (DC) withstand voltage methods should not be used. DC withstand voltage tests can be performed on the paper-insulated cable. Due to the presence of insulation defects inside the XLPE cable, there is a high risk of partial discharge. Applying DC voltage at this time will further accelerate insulation aging and cause electrical treeing. The structure of the XLPE power cable has a "memory effect." They tend to accumulate residual charges under DC voltage, which takes a long time to dissipate. If the cables are put into operation before the charges are completely dissipated, the DC bias will be superimposed on the AC voltage, leading to insulation damage. Insulation breakdown points detected during DC withstand voltage tests are often difficult to breakdown under AC operating conditions, while points prone to insulation breakdown in AC conditions are frequently undetectable in DC withstand voltage tests.

Therefore, even the XLPE power cable that has passed DC withstand voltage tests often experiences insulation breakdown accidents during operation. Therefore, 0.1 Hz very low frequency (VLF) test equipment should be used for the XLPE cable. DC high voltage tests are half-wave tests and cannot effectively detect insulation defects such as water trees in cross-linked polyethylene (XLPE) insulation. It can also lead to insulation breakdown shortly after the high-voltage cable is put into operation despite passing the DC test, thus failing to achieve the desired testing purpose effectively. Therefore, for the XLPE cable, it is more common to use variable frequency series resonant withstand voltage test equipment for AC withstand voltage tests on capacitive test samples. This method has the advantages of good power frequency equivalence, high fault detection rate, minimal damage to the test sample, small size, and light weight of the test equipment, making it suitable for on-site transportation.

(1) After the DC test, it is necessary to discharge the residual charges. The discharge process consists of grounding through a resistor and then direct grounding. For large-capacity test specimens such as large capacitors, large motors, long cables, etc., the discharge time should not be less than 5 minutes to ensure sufficient discharge of charges on the test specimen. Each stage of the multi-stage voltage-boosting rectifier device should be fully discharged before changing the wiring.

(2) If there is an abnormal leakage current during the measurement process, methods such as drying or shielding can be used to assess the insulation status of the equipment. Shielded wires should be used for high-voltage leads to prevent leakage current from affecting the results.

High-voltage leads should not produce corona discharge, and microammeters should be used to measure the high-voltage end. If a sharp increase in leakage current, insulation burning smell, smoke, noise, or other abnormal phenomena are observed during the test, the voltage should be immediately reduced, the power should be disconnected, the test should be stopped, and the windings should be grounded and discharged before conducting further inspections.

(3) The test voltage for the graded insulation transformer should be based on the voltage level of the tested winding but should not exceed the withstand voltage level of the neutral point insulation. The leakage current of the 500 kV transformer is generally not greater than 30 µA.

(4) When encountering abnormal data during measurement, the following factors can be investigated: test wiring, the installation position of the microammeter, ambient temperature and humidity, internal temperature of the test specimen, degree of dirt on the surface of the test specimen, high-voltage leads, and the influence of surrounding electric fields or power grids.

(5) When the zinc oxide arrester has multiple sections, each section must meet the requirements. Single microammeter or dual microammeter wiring methods can be used.

Module 4 Training

What is the difference between DC withstand voltage test and AC withstand voltage test?

___
___
___

Worksheet 3　Dielectric Loss Test

Module 1　Operating Worksheet: Dielectric Loss Test

(Ⅰ) Test Name and Instrument	(Ⅱ) Test Objects
Dielectric loss measurement **Anti-interference dielectric loss automatic tester**	Dielectric loss and capacitance measurement of equipment such as transformer, reactor, bushing, capacitor, lighting arrester, etc. The test parameters include dielectric loss tangent and capacitance
(Ⅲ) Test Purposes	(Ⅳ) Measurement Steps
(1) Detect overall insulation moisture, deterioration, as well as defects in small-sized tested equipment, whether they are penetrated or not. (2) Detect defects of permeable conductive channels in the dielectric. (3) Detect defects such as insulation bubbles and aging	(1) Connect the test lines as required, connect the high-voltage leads in order, and take grounding safety measures. (2) Power on, select the high-voltage generation method (internal connection or external connection), connect the wiring method (positive connection, reverse connection, etc.), and select the high voltage value. (3) Press the high-voltage allowance key, then press test key to detect the dielectric loss and capacitance. (4) Save the data and turn off the power. (5) Discharge
(Ⅴ) Precautions	(Ⅵ) Technical Standards
(1) Ground and discharge the test object before and after the test. (2) Select the correct connection method (positive or reverse) according to whether the equipment is grounded. (3) Dielectric loss tests at high voltage can be lethal; the instrument panel should be reliably grounded. (4) High-voltage leads should be suspended overhead	The dielectric loss at 20 °C should not exceed 0.8%, and the comparison with historical data should not exceed 30%

Continued

(Ⅶ) Result Judgment	(Ⅷ) Digital resources
(1) When the voltage level of the transformer is 5 kV or above and the capacity is 8,000 kVA or above, the dielectric loss tangent value tanδ should be measured. (2) The measured tanδ value of the winding should not exceed 130% of the product's factory test value. (3) When the temperature during measurement does not match the product's factory test temperature, conversion to the same temperature should be performed for comparison	Measurement of transformer dielectric loss tangent (tanδ) at different frequencies Transformer winding main transformer winding dielectric loss test

Module 2 Follow Me

Ⅰ. Measurement Purpose

When the insulation resistance of the specimen is low and the leakage current is high, it is not sufficient to determine whether the specimen's overall insulation is unqualified. Further judgment can be made through measurement of dielectric loss. If the electrical equipment is excessively large in size, different components can be measured separately according to their structures.

Dielectric loss factor measurement is a highly sensitive test, it can detect overall moisture, degradation, and defects of small-sized specimens with or without penetration. The tanδ value of the tested winding should not exceed 130% of the product's factory test value.

Ⅱ. Technical Standards

The "Test Procedures" specify that when measuring the tangent of the dielectric loss angle (tanδ) for the winding and its bushing, the following criteria should be met: For transformers with voltage ratings of 5 kV and above, and capacities of 8,000 kVA and above, the tangent of the dielectric loss angle (tanδ) should be measured. Qualified standards for dielectric loss as shown in table 2-7.

Table 2-7 Qualified Standards for Dielectric Loss

ID	Power Equipment	Qualified Standards	Note
1	Transformer	Dielectric loss at 20 °C should not exceed 0.8%. Comparison with historical data should not exceed 30%	
2	Instrument transformer	(1) After major repairs, the dielectric loss factor should not exceed 3.0. During operation, it should not exceed 3.5. (2) After major repairs, the dielectric loss factor for the secondary winding should not exceed 2.0. During operation, it should not exceed 2.5. (3) There should not be significant changes when compared with historical data	

ID	Power Equipment	Qualified Standards	Note
3	Circuit breaker, disconnect switch	(1) For the oil circuit breaker rated at 110 kV and below, the dielectric loss should be less than 2%. For other types, it should be less than 1%. (2) References may be made to manufacturer's technical standards for other criteria	
4	Reactor	For reactors rated at 35 kV and below, the dielectric loss factor should be 3.5. For 66 kV, it should be 2.5	This criterion applies only to the oil-immersed iron core reactor with a capacity of 800 kvar or above
5	Capacitor	The $\tan\delta$ value for the capacitor under 10 kV should not exceed the following limits: (1) Oil-paper insulation: 0.005. (2) Film-paper composite insulation: 0.002	

III. Measurement Principle

$\tan\delta$ is a characteristic parameter that reflects the magnitude of insulation material loss, and it is particularly sensitive to overall defects in low-capacity power equipment. However, for high-capacity and multi-component devices such as the transformer, cable, and generator, if the defects in the insulation are not distributed uniformly but rather concentrated locally, and if the insulation volume is larger or the proportion of concentrated defects is smaller, the ratio of localized defect dielectric loss to the total insulation dielectric loss becomes smaller. In such cases, $\tan\delta$ measurement may be not sensitive enough. Therefore, when measuring the $\tan\delta$ of large capacity or multi-component power equipment, the equipment should be disassembled for testing, and the $\tan\delta$ values of each component should be measured separately to determine if there are localized defects in the tested specimen.

Figure 2-9 shows the functional structure of a high-voltage dielectric loss factor test instrument.

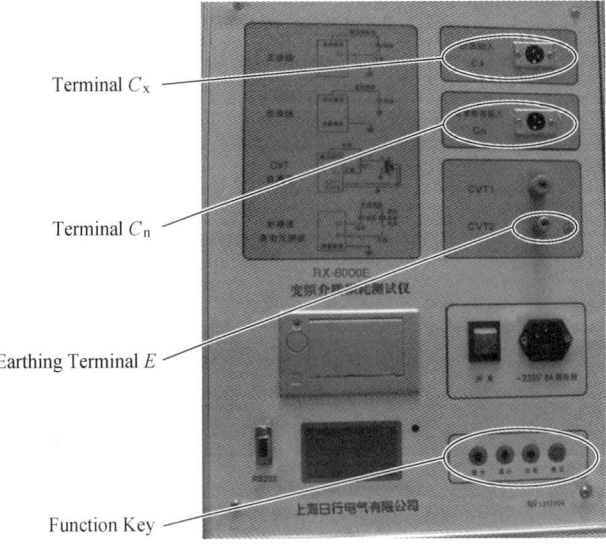

Figure 2-9 Functional structure diagram of high voltage dielectric loss instrument

IV. Testing Principle

The wiring principle of the high-voltage schering bridge is shown in Figure 2-10. When the bridge is balanced, no current flows through the ammeter.

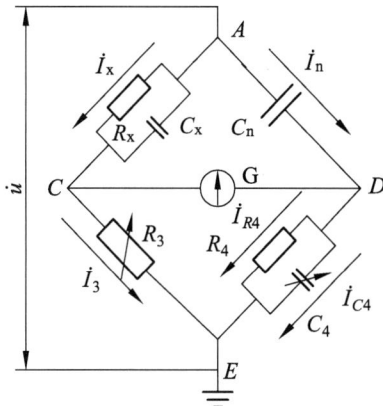

Figure 2-10 Working principle of high-voltage Schering bridge

C_x and R_x: the capacitance and resistance (in series equivalent circuit) of the test specimen.

R_3: adjustable resistor.

G: galvanometer.

R_4: fixed resistor.

C_n: standard capacitor (50 ± 1) pF.

C_4: adjustable capacitor.

R: protective resistor.

Module 3 Workshop

I. Wiring for measurement

There are two methods for measuring the dielectric dissipation factor value $\tan\delta$: positive wiring and negative wiring. Figure 2-11 shows the high and low-voltage wiring diagram of a transformer winding, and Figure 2-12 shows the wiring diagram for measuring the dielectric dissipation factor of a transformer bushing.

1. Positive Wiring Method

Wiring Characteristics: The test specimen PX is insulated from the ground (which is sometimes not easy to achieve in the field).The test specimen is at high voltage, and one end of the bridge is grounded. In the positive wiring method, the high-voltage electrode of the standard capacitor, the high-voltage terminal of the test specimen, and the high-voltage electrode of the voltage-boosting transformer all have dangerous voltages. Therefore, the bridge measurement part must be reliably grounded, and the testers should keep a safe distance. Figure 2-13(a) shows the schematic diagram of the positive wiring method.

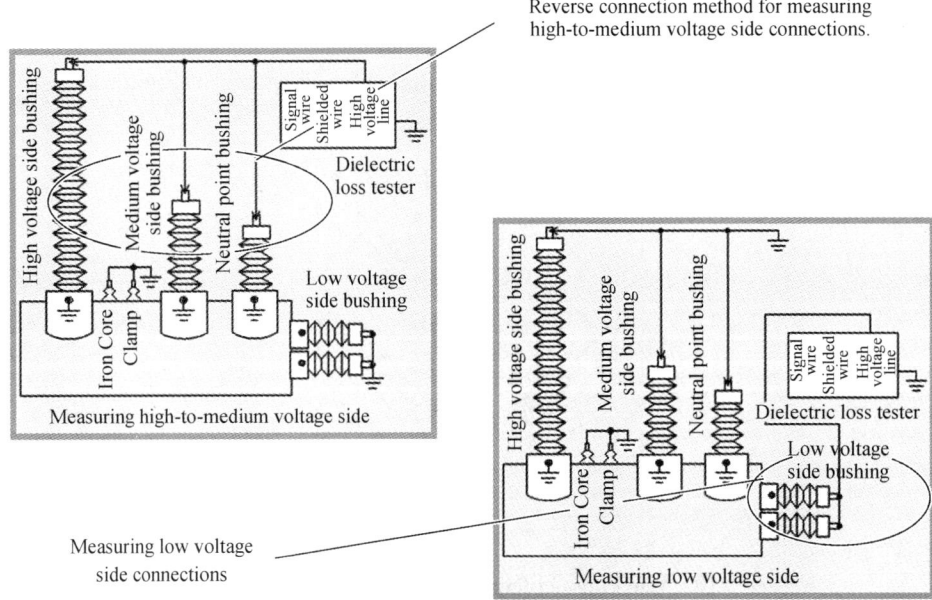

Figure 2-11 Wiring diagram for transformer winding high and low voltage connections

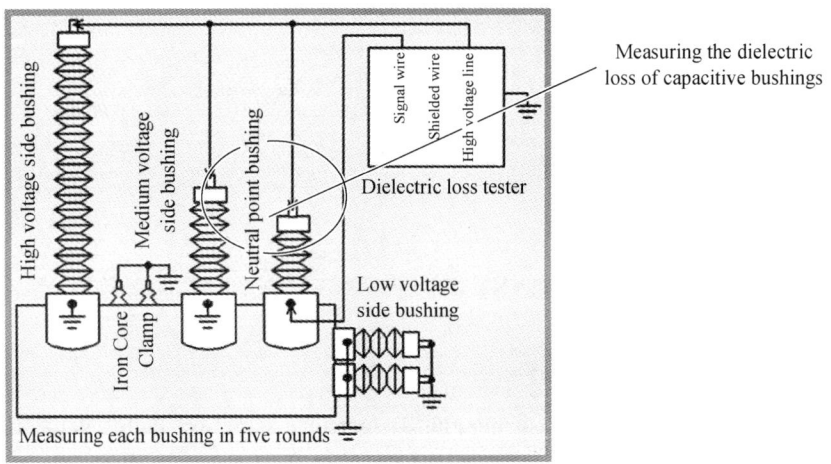

Figure 2-12 Wiring diagram for transformer bushing dielectric loss factor measurement

2. Negative Wiring Method

Wiring Characteristics: One end of the test specimen PX is grounded, and the bridge is at a high voltage potential.

The housing of the standard capacitor carries a high voltage, so it is important to ensure that the bridge is well grounded and to avoid reaching behind the bridge during the test. It is necessary to insulate the housing from the ground and maintain a certain distance from the grounding wire. Caution should be exercised during operation and reading. Figure 2-13(b) shows the schematic diagram of the negative wiring method, and Figure 2-14 illustrates the wiring and measurement process of the high-voltage schering bridge in the negative wiring method.

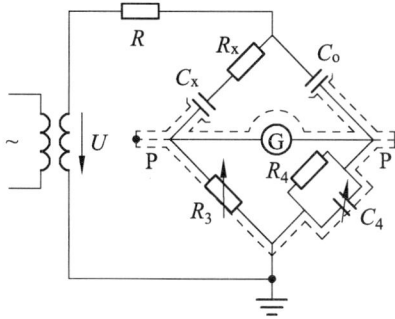

(a) Wiring diagram for high-voltage schering bridge positive wiring method

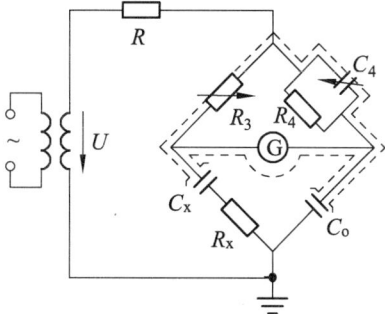

(b) Wiring diagram for high-voltage schering bridge negative wiring method

Figure 2-13　Wiring for measurement

Dielectric loss test for transformer windings

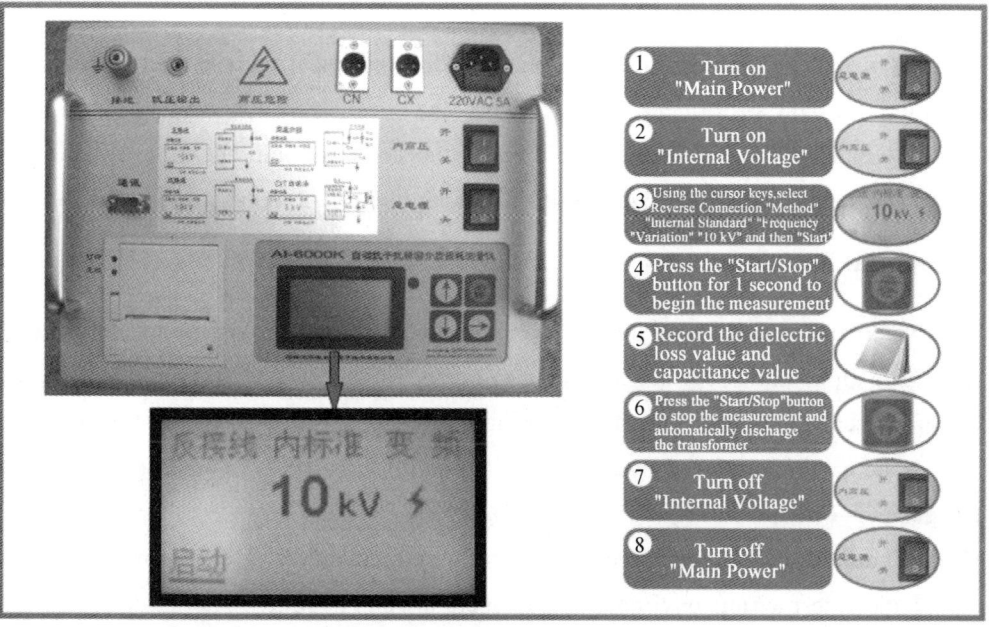

Figure 2-14　Wiring and measurement process of the high-voltage Schering bridge in negative wiring method

Before use, it is necessary to ensure that the bridge measurement part and grounding terminal of the instrument are reliably grounded. In the positive wiring method, the high-voltage electrode of the standard capacitor, the high-voltage terminal of the test specimen, and the high-voltage electrode of the voltage-boosting transformer all carry dangerous voltages. The interconnections between these terminals should be kept open, and testers should keep a safe distance. Before approaching the measurement system, connecting or disconnecting wires, and charging the measurement unit's power supply, make sure that all measurement power sources have been switched off. It is also important to pay attention to the safety of the low-voltage power supply.

(1) Only when the "Internal High Voltage Allowed" key of the instrument is not pressed, it is safe to touch the back panel of the instrument and the measurement cables connected to the test specimen. When the "Internal High Voltage Allowed" key is pressed, the buzzer will sound as a warning signal.

(2) While the instrument is in measurement mode, all buttons except the "Start" button should not be operated. However, the measurement status can be exited using the "Start" button.

(3) When measuring a non-grounded test specimen (positive wiring method), the "HV" terminal is at a high voltage with respect to ground. When measuring a grounded test specimen (negative wiring method), the "C_x" terminal is at a high voltage with respect to ground. The red and blue cables provided with the instrument are high-voltage shielded cables, which can be laid along the ground when in use. However, the outer shielding of the cable must be connected to a dedicated grounding terminal.

(4) Do not replace the fuse tube with one that does not meet the specified value to prevent damage to the internal transformer.

The back panel of the instrument should be kept clean and should not be touched with hands. If there are any stains on the back panel, use a dry cloth to clean it and ensure good insulation.

II. Special Notice

(1) Due to the formation of low-resistance circuits on the surface of insulating materials caused by high air humidity, surface leakage may occur, which can interfere with the measurement of $\tan\delta$ of power equipment. Therefore, the humidity during measurement should not exceed 80%. This issue can be addressed by applying silicone oil, paraffin wax, or drying the surface of the test specimen ceramic bottle.

(2) If abnormal data is obtained during the measurement, it is necessary to examine the following factors: test wiring, environmental temperature and humidity, internal temperature of the test specimen, extent of surface contamination on the test specimen, position and length of the high-voltage leads, and external electric field or power grid interference. When $\tan\delta$ is 0 or a

negative value, it is necessary to check if the air humidity is too high and causing the formation of a water film, whether the connection between the high-voltage lead and the capacitance screen is reliable, and if the flange of the bushing is properly grounded. These issues can be addressed by cleaning the surface of the test sample, improving the connections or grounding, and increasing the diameter of the high-voltage lead, among other methods.

(3) Since the oil temperature of the transformer reaches 70 °C–90 °C during normal operation, and the tanδ of insulating oil increases with temperature, this change is more obvious in aging and damp transformer oil. Therefore, when measuring the tanδ of insulating oil, the oil is generally heated to around 90 °C before measurement. For the oil-immersed or oil-paper capacitor-type current transformer, if there is water and impurities in the insulation, they can also be measured after the insulation is heated up using a short-circuit method. If the tanδ increases significantly at high temperatures, it can be considered that there is an insulation defect in the equipment.

(4) When measuring the tanδ of a transformer, the iron core must be grounded. If the iron core is not grounded, the insulating material such as hardboard between the transformer windings and the casing becomes connected, leading to an increase in the measured insulation value and a potential failure to meet the required tanδ specifications.

(5) The presence of partial discharge defects in electrical equipment can also cause a decrease in tanδ. Therefore, it is important that the tanδ of qualified power equipment does not show significant increases or decreases compared to previous measurements.

(6) Since positive wiring has good anti-interference capabilities, it is advisable to use the positive connection method during testing and use the data obtained for insulation condition assessment. A comparison of data can also be made between the results obtained from positive and negative wiring.

(7) When measuring tanδ, it is also advisable to measure the capacitance of the test specimen simultaneously. If there are noticeable changes in capacitance compared to previous measurements, it is necessary to examine the internal insulation for potential defects or perform a thorough inspection. For instance, when the insulation of a transformer becomes damp, the capacitance value may decrease at low temperatures and increase when measured at high temperatures.

(8) When measuring the capacitive high-voltage bushing, it is important to consider their small capacitance and their vulnerability to the impact of surrounding electric fields and ground. This can result in measurement dispersion errors when they are placed in different positions. To minimize these errors, it is recommended to position the bushing vertically on securely grounded bushing stands, rather than suspending them horizontally with insulating ropes. The correct vertical placement of the capacitive high-voltage bushing on a bushing stand is depicted in Figure 2-15.

Figure 2-15 The Capacitive high-voltage bushing placed on the bushing frame vertically

Module 4 Training

How can the quality of equipment be determined based on the results of dielectric loss testing?

Worksheet 4 AC Voltage Withstand and Series Resonance Withstand Test

Module 1 Operating Worksheet: Withstand Voltage Test

(Ⅰ) Test Name and Instrument	(Ⅱ) Test Objects
Power frequency withstand voltage test **Power frequency AC withstand voltage equipment**	Insulation strength test of high voltage equipment such as transformer, generator, motor, instrument transformer, lightning arrester, switch, insulator, capacitor, power cable, etc. under specified voltage
(Ⅲ) Test Purpose	(Ⅳ) Test Steps
(1) The most direct and effective method to test the insulation strength of power equipment. (2) Test the insulation margin of the equipment. (3) Check if the equipment meets the requirements for safe operation	(1) Connect the test lines as required, connect the high-voltage leads in order, and take appropriate grounding safety measures. (2) In the first step, do not connect the test object. Based on the preset protective voltage value, set the distance between the protective spheres. The protective sphere gap protection voltage should be 1.1 to 1.2 times the test voltage. (3) Turn on the power, reset the voltage device to zero, press the high-voltage permit key, and apply high voltage until the protective sphere gap breaks down. (4) Quickly reduce the voltage to zero, turn off the power, and discharge. (5) Connect the test object in parallel, with the high-voltage leads suspended. (6) Turn on the power, press the high-voltage permit key, and apply high voltage to the withstand voltage value. Within 60 seconds, there should be no breakdown or discharge. Record the voltage value and then reduce it to zero. (7) Discharge

(V) Precautions	(VI) Technical Standards
(1) Ground the test object before and after the test and discharge it thoroughly. (2) The high voltage used in the withstand voltage test can be lethal, so the operating table must be reliably grounded. (3) Adjust the protection gap size before connecting the test sample for the high-voltage test. (4) Before applying voltage, carefully check whether the wiring is correct and maintain a sufficient safety distance. (5) The high-voltage leads must be suspended, and the voltage increasing process should be called out to each other. (6) After the test, restore the various connections, tap shielding ends, and cover plates of the test object	(1) Before conducting high-voltage tests, perform low-voltage tests first. After passing non-destructive tests such as insulation resistance, proceed with AC withstand voltage tests and partial discharge destructive tests. (2) The withstand voltage test is a destructive test. Therefore, before and after the test, conduct insulation resistance tests to ensure that the difference in insulation resistance does not exceed 30%. This helps determine whether there are any changes in the insulation of the test sample before and after the withstand voltage test
(VII) Result Judgment	(VIII) Digital Resources
(1) If no destructive discharge occurs during the test, the withstand voltage test is considered successful. (2) During the voltage increasing, if significant fluctuations in ammeter pointer, smoke, burning smell, flashover, or breakdown of the test sample are observed, the AC withstand voltage test is considered unsuccessful. Immediately stop test and reduce the voltage and cut off the power. (3) If the insulation resistance after the withstand voltage test decreases by 30% compared to before, inspect the test sample to determine its qualification	AC withstand voltage test of circuit breaker Variable frequency resonant voltage withstand test device

Module 2 Follow Me

I. Overview of AC Withstand Voltage Test

Due to factors such as moisture, insulation aging, and damage, the insulation performance of electrical equipment may deteriorate. The AC withstand voltage test is the most direct and effective method to evaluate the insulation adequacy of electrical equipment. The purpose of the test is to assess the installation quality and insulation strength of the electrical equipment. Generally, the withstand voltage test is conducted during equipment handover, major repairs, and annually after passing the insulation preventive test to prevent insulation accidents.

Since the AC withstand voltage test is a destructive test, prior tests such as insulation resistance, DC leakage current, and dielectric loss must be performed on the test specimen. Only when the test results are normal can the AC withstand voltage test be conducted. If poor insulation conditions such as moisture and local defects are found in the equipment, they should be dealed well before conducting the withstand voltage test to avoid unnecessary insulation breakdown. Newly installed oil-insulated equipment, such as transformers and circuit breakers, also require a waiting period of at least 48 hours for all gas bubbles to escape from the oil before the test can be performed. If it is gas-insulated electrical equipment, the test should be conducted under the lowest allowable gas pressure to easily detect internal insulation defects.

The withstand voltage test can be divided into the following categories: power frequency withstand voltage test, induced voltage withstand voltage test, and impulse voltage test. The power frequency withstand voltage test includes conventional power frequency AC tests, power frequency resonance withstand voltage tests, and 0.1 Hz ultra-low frequency withstand voltage tests.

II. Power Frequency AC Withstand Voltage Test

The power frequency AC withstand voltage test equipment is shown in Figure 2-16.

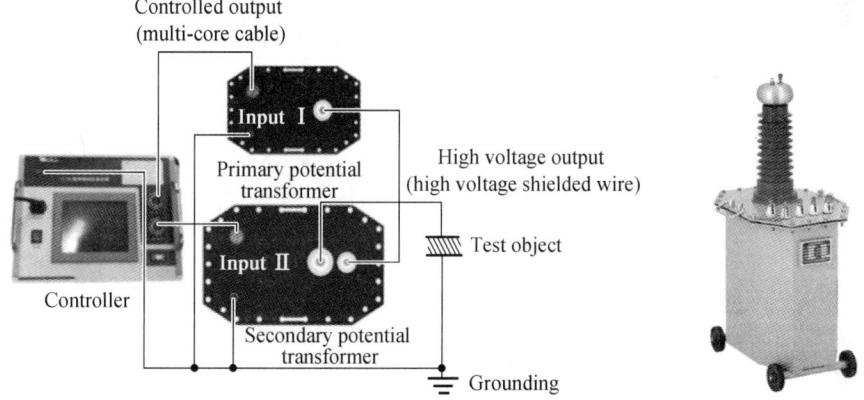

Figure 2-16　Power frequency AC withstand voltage test equipment

III. Resonant Withstand Voltage Test

1. Series Resonant Withstand Voltage Principle

When conducting withstand voltage tests on high-voltage capacitive test objects such as the generator, transformer, power cable, GIS (Gas Insulated Switchgear), switch, bushing, and transformer, a series or parallel resonant device can be employed. In the case of series resonant testing, an excitation transformer and reactor are utilized. The series resonant power source takes advantage of the resonance between the reactor and the capacitance of the test object to generate high voltage and large current. This results in a significant reduction in the required power capacity of the power source to only $1/Q$ of the test capacity. It eliminates the need for bulky high-power voltage regulators and high-power frequency test transformers. The weight and volume of the series resonant equipment are generally 1/10 to 1/30 of conventional test setups.

In a series resonant state, when the insulation weak points of the test object breakdown, the circuit immediately becomes detuned, and the circuit current rapidly decreases to 1/Q of the normal test current. This method effectively detects the insulation weak points while preventing excessive current and burn damage at the short-circuit fault points. Additionally, it avoids the occurrence of overvoltage during the recovery period.

The capacity of test equipment required for conducting AC withstand voltage tests on high-voltage, high-capacity equipment has been increasing. Conventional power frequency withstand voltage methods often fail to meet the requirements of on-site testing. Therefore, series resonant testing methods are widely adopted for on-site testing. The series resonant testing equipment is characterized by its small size, low voltage of the test power source, low power capacity, and good waveform of the test voltage. As a result, series resonant testing is widely applied in on-site AC withstand voltage, induced withstand voltage, and partial discharge tests for high-voltage, high-capacity power equipment such as cable, Gas Insulated Switchgear (GIS), large generator, transformer, and capacitor.

The series resonant withstand voltage test applies the basic principle of resonance circuits. According to different adjustment methods, it can be divided into inductance adjustment, capacitance adjustment, and frequency conversion adjustment types. Through comparisons of the three types of series resonant devices in practical applications, it was found that frequency conversion series resonant devices are more suitable for on-site needs. During on-site AC equipment tests, the frequency conversion series resonant device can fulfill various AC withstand voltage tests. Refer to Figure 2-17 for details. The device has three operating modes: fully automatic mode, manual mode, and automatic tuning manual boosting mode.

Figure 2-17　Frequency conversion series resonant withstand voltage test device

There are commonly used methods for adjusting the series resonant withstand voltage test in the field: inductance adjustment and frequency conversion adjustment. Regardless of the method used, the goal is to adjust the high-voltage circuit so that the impedance equals the admittance, i. e., $L=1/C$. By doing so, a lower voltage can be applied to the intermediate transformer, while generating a high voltage QU ($Q=L/R=1/\omega CR$, where ω is the angular frequency and C is the capacitance) across the test specimen.

Whether a sufficient high voltage can be generated across the test specimen depends on the following three conditions:

(1) Whether the total impedance of the high-voltage circuit equals the total admittance.

(2) Whether the quality factor of the high-voltage circuit is sufficiently large.

(3) Whether the voltage U applied by the intermediate transformer is large enough.

The most important component of the inductance adjustment device is the adjustable iron core reactor, which adjusts the inductance of the reactor by changing the air gap of the iron core. The advantage of this device is that the inductance of the reactor can be made very large, allowing for high-capacitance withstand voltage tests on large-capacity specimens. Moreover, the size of the inductance can be linearly adjusted, allowing the series circuit to reach the resonance point precisely according to the capacitance of the specimen. By initially applying a low voltage to adjust the inductance of the reactor and induce resonance in the series circuit, the voltage can then be increased while adjusting the inductance of the reactor until resonance occurs at the test voltage. This approach ensures safety and achieves complete resonance in the circuit. The drawback is that the manufacturing of the reactor is complex and adds weight to the device.

The frequency conversion device relies on a high-power frequency converter to adjust the power supply frequency, allowing the circuit to reach the resonance point. The inductance of the reactor used in this device is fixed, while the frequency of the test power supply varies depending on the capacitance. The frequency conversion series resonant test device utilizes the principle of series resonance, exciting the series resonance circuit using an excitation transformer, and adjusting the output frequency of the frequency converter to match the circuit inductance L and specimen capacitance C.

In series resonance, the resonant voltage is exactly applied to the specimen. The advantages include precise resonance point achievement, simple construction of the reactor, lightweight, and high-quality factor of the circuit. The drawback is the requirement for a high-power power supply and high stability of the power supply voltage and frequency. The frequency range of the frequency conversion device needs to be larger than that of the inductance adjustment device. With the advancement of electronic equipment, the frequency conversion method is generally preferred nowadays.

The frequency conversion series resonance test utilizes frequency modulation and voltage regulation. It achieves capacitive resonance by utilizing the inductance of the reactor and the capacitance of the test specimen. The excitation transformer supplies the excitation power to the resonance circuit. By adjusting the inductance or changing the output frequency of the power supply, the reactive impedance and capacitive impedance in the circuit can be made equal, resulting in a resonant state in the circuit. In this state, the reactive power in the circuit tends to zero, and the circuit current is maximized and in phase with the input voltage. This enables the inductance or capacitor terminals to obtain a voltage that is Q times higher than the excitation voltage. This method is particularly suitable for AC withstand testing circuits of large-capacity and high-voltage test specimens. As shown in table 2-8.

Table 2-8　Selection of resonant test voltage and test specimens

ID	Test Voltage	Test Object Withstand Voltage	Reactor (45 kVA/45 kV, 6 sections)	Excitation Transformer Output Voltage Selection (4 kV, 8 kV, 16 kV)	Test Time
1	22 kV	Meet the AC withstand test of the 3 km 10 kV/300 mm² cable	a six-section reactor is used in parallel (three of which are grounded)	4 kV	5 min
2	35 kV	Meet the AC withstand test of the 10 kV/20,000 kVA transformer and instrument transformer	a two-section reactor is used in series with compensation capacitors	4 kV	1 min
3	42 kV	Meet the AC withstand test of the 10 kV switch, instrument transformer and busbar	a one-section reactor is used in series	4 kV	1 min
4	52 kV	To meet the AC withstand test of the 1 km 35 kV/300 mm² cable	Two series reactors and three parallel reactors are used	4 kV	60 min
5	68 kV	Meet the AC withstand test of the 35 kV/31,500 kVA transformer	a three-section reactor is used in series with compensation capacitors	8 kV	1 min
6	95 kV	Meet the AC withstand test of the 35 kV switch, instrument transformer, and busbar	a three-section reactor is used in series	16 kV	1 min
7	128 kV	Meet the AC withstand test of the 0.2 km 110 kV/300 mm² cable	Three series reactors and two parallel reactors are used	4 kV	60 min
8	160 kV	Meet the AC withstand test of the fully insulated main transformer and instrument transformer of 110 kV/80,000 kVA	a four-section reactor is used in series	4 kV	1 min
9	184 kV	Meet the AC withstand test of the 110 kV switch	a six-section reactor is used in series	16 kV	1 min
10	265 kV	Meet the AC withstand test of the GIS switch of 110 kV and below	a six-section reactor is used in series	16 kV	1 min

Special note: During the handover test, the withstand voltage test time for 10 kV cables is 5 minutes, and for cables of 35 kV and above, the withstand voltage test time is 60 minutes.

2. Resonant Withstand Voltage Device

The characteristic of a frequency-variable series resonance AC test device is that the tuning reactor has a small weight and simple structure, making it more suitable for on-site testing of high-capacity equipment. The wiring principle of the frequency-variable series resonance is shown in Figure 2-18.

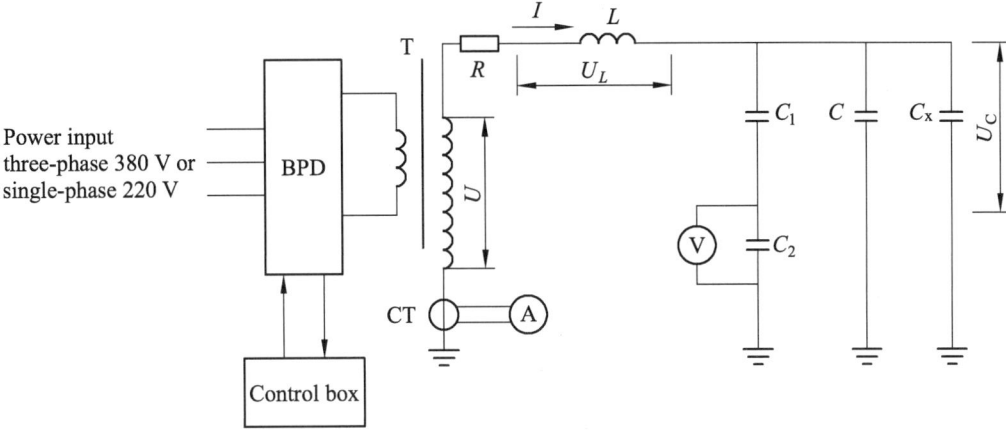

BPD–variable frequency power supply; T–excitation transformer; C_x–equivalent capacitance of test sample; C_1、C_2–capacitive voltage divider; C–compensation capacitor; L resonance reactor; R–equivalent resistance of the test circuit; U–excitation voltage; U_L–reactor voltage; U_C–test object voltage; I–test circuit current; CT–current sampling.

Figure 2-18　Frequency-variable series resonance principle

The series resonance AC withstand test device mainly consists of a variable frequency power supply (host), an excitation transformer T, a resonance reactor L, and capacitor voltage dividers C_1 and C_2, as shown in Figure 2-19.

Figure 2-19　Series resonant withstand voltage

(1) Variable frequency power supply.

A variable frequency power supply is a power source with adjustable frequency within a certain range, as shown in Figure 2-20. The output power of the variable frequency power supply is generally greater than the output capacity of the excitation transformer, and the frequency adjustment range is 20 to 300 Hz.

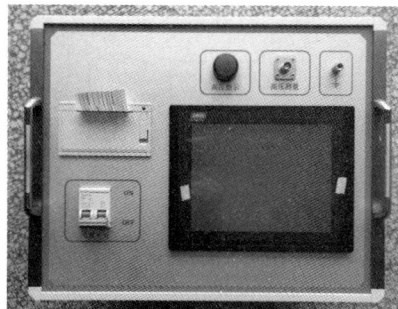

Figure 2-20 Variable frequency power supply

(2) Excitation transformer T.

The excitation transformer T provides the excitation power supply for the resonance circuit. It excites the series resonance circuit. By adjusting the output frequency of the variable frequency controller, the circuit inductance L and the test object capacitance C are series resonated. The resonance voltage is then applied to the test sample. When wiring, the high-voltage tail X of the excitation transformer must be grounded.

(3) Resonant Reactor L.

The resonance reactor is used to resonate with the capacitive equipment in the test circuit and obtain high voltage. The capacity of the resonance reactor can be relatively large to meet the requirements of the test. The rated voltage of the resonance reactor should meet the requirements of the test voltage, and its rated capacity should meet the requirements of the test capacity. The reactor can be a fixed single-stage reactor or an adjustable reactor adjusted directly through knobs on the instrument. The single-stage reactor can be flexibly combined in series or parallel, making it convenient to meet the needs of on-site testing.

(4) Capacitor Voltage Dividers C_1 and C_2.

The voltage dividers are connected in parallel to the test object. They are used to measure the resonance voltage on the test sample and provide overvoltage protection, as shown in Figure 2-21. When calculating the parameters of the resonance system, the capacitance of the capacitor voltage dividers should be taken into consideration.

(5) Capacitor Compensator C.

The capacitor compensator is used to compensate for the inductance of the test circuit, ensuring that the test circuit meets the resonance conditions and test requirements. The rated voltage of the capacitor compensator should meet the test requirements.

Figure 2-21 Capacitor Voltage Dividers

In the power frequency state, variable inductors generate resonant current within a specific range when combined with variable reactance, fulfilling the test requirements.

3. Resonant Withstand Voltage Wiring and Precautions

During wiring, the resonance reactor, voltage dividers, excitation transformers, and other components of the variable frequency series resonance test equipment should be placed as close as possible to the test sample to minimize the length of the connections. Additionally, a short-circuit grounding wire should be used for grounding. During the test, irrelevant personnel should not approach the variable frequency series resonance equipment without permission. Serial resonance AC tests should not be conducted outdoors on rainy days. Interference in partial discharge tests during series resonance should be avoided.

Strengthen the fast joint for high-voltage leads. The length of the high-voltage leads is usually around 10 meters, and they are externally covered with metal hoses to ensure a uniform surface electric field for the leads. The two ends of the metal hoses should be reliably connected.

Main characteristics of series resonance systems:

(1) Wide applicability, small size, light weight, large testing capacity, and high test voltage.

(2) High safety and reliability, simple operation, and good test equivalency.

(3) The series resonance device has a high impedance for higher harmonic components in the circuit, resulting in a good voltage waveform for the tested specimen. In the event of flashover breakdown during withstand voltage testing, the resonance conditions are lost, the high voltage immediately disappears, and the arc is extinguished, thereby protecting the tested specimen.

Ⅳ. Other Withstand Voltage Tests

1. 0.1 Hz ultra-low Frequency Withstand Voltage Test

The 0.1 Hz ultra-low frequency withstand voltage test, still considered an AC withstand voltage test, can effectively detect defects in capacitive equipment and inspect the operational and installation quality of equipment such as generator, transformer, and rubber-insulating power cable. It assesses the insulation strength of the main insulation of generators and transformers, cable terminations, and intermediate joints, and can conduct testing sensitively, as shown in Figure 2-22.

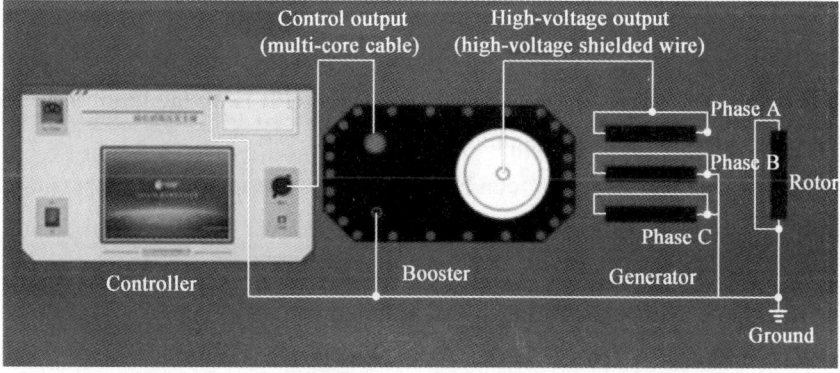

Figure 2-22   0.1 Hz ultra-low frequency withstand voltage test

The 0.1 Hz ultra-low frequency withstand voltage test is more effective in detecting defects in the insulation at the ends of generators compared to the power frequency withstand voltage test. The reason is that under power frequency voltage, the capacitance current flowing out from the busbar causes a significant voltage drop when passing through the semiconductor corona protection layer outside the insulation, resulting in a reduced voltage on the busbar insulation at the ends. In the case of ultra-low frequency, this capacitance current is greatly reduced, and the voltage drop on the semiconductor corona protection layer is also significantly reduced. As a result, the voltage on the insulation at the ends is higher, making it easier to detect defects.

Wiring method: The test should be conducted phase by phase, applying voltage to the phase under test and short-circuiting the non-tested phases to the ground. The test duration for commissioning test is 60 minutes, and for preventive test, it is 15 minutes.

2. Induced overvoltage withstand test

The induced overvoltage withstand test is often applied to electrical equipment such as the transformer and the electromagnetic potential transformer. The method used involves applying a high voltage to the primary side through secondary side voltage boosting. This approach can be used to test the main insulation of the device under test and also effectively inspect longitudinal insulation, often using multiples of 100–400 Hz.

Taking transformer withstand voltage test as an example, the power frequency withstand voltage test evaluates the electrical strength of the main insulation of the transformer windings. This includes insulation between the high, medium, and low windings and the grounded parts such as the iron core and oil tank. Induced overvoltage withstand tests can focus on the main insulation and the longitudinal insulation of the windings (insulation between turns, layers, and sections), generally using double-frequency (100 Hz) or triple-frequency (150 Hz).The excitation voltage frequency of the induced voltage test should typically be three times the frequency (150 Hz) and should not exceed 400 Hz, with the withstand time being $60 \times 100/f(s)$ and the duration not less than 20 s.

3. Impulse Voltage Test

The impulse voltage test is mainly to test the ability of the test object to withstand the operating wave overvoltage and the atmospheric overvoltage. It is divided into the operating wave impulse voltage test and the lightning impulse voltage test.

The impulse voltage test is mainly conducted to assess the insulation withstand capability of the DUT (Device Under Test) against operating impulse over voltages and atmospheric over voltages. It can be divided into operating impulse voltage tests and lightning impulse voltage tests. Regardless of the type of withstand voltage test, suitable test equipment should be selected based on the nameplate parameters of the DUT, test voltage magnitude, and the existing test equipment conditions, in order to meet the requirements of the test.

During on-site arrangement and wiring, pay attention to maintaining a sufficient safety distance for both high voltage to ground and test personnel. The high-voltage leads should be securely and as short as possible, and the non-DUT phase and equipment shell should be reliably grounded. After completing the wiring, the person in charge should carefully inspect the capacity, range, and position of the test equipment, ensuring that the voltage regulator indicator is at zero position and that all connections are correct and error-free.

Module 3 Workshop

Ⅰ. Test preparation

1. Understand the on-site situation of the DUT and the test conditions

Conduct site survey and review relevant technical documents, including the annual test data and related regulations of the equipment, in order to understand the operation and defects of the equipment.

2. Test instrument and equipment preparation

Select appropriate test transformer and control panel, series-resonant high-voltage withstand device, protective resistor, sphere gap, voltage divider, digital multimeter, megohmmeter, discharge rods, insulation operating pole, grounding wire, high-voltage cable, multimeter, Hygrometer, electrical common tool, safety belt, helmet, temporary safety barrier, sign, etc. Check the validity period of the calibration certificates for the test instruments, equipment, and insulation tools.

3. Test transformer inspection

Use a 2.5 kV megohmmeter to check the insulation resistance of each winding to the casing and ground. Check if the high-voltage coil circuit is connected by measuring the resistance value between the high-voltage terminal using a multimeter with a 1 kV range. The pointer should clearly deflect towards the direction of lower resistance.

4. Issue work permits and implement on-site safety and technical measures

Explain the work content, energized areas, on-site safety measures, and potential hazards to other test personnel, clearly assign tasks and specify the test procedures.

Ⅱ. Power frequency withstand voltage test operation

The power frequency withstand voltage test circuit consists of a test transformer, voltage regulator equipment, measurement circuit, control, and protection circuit, etc.

1. Test wiring

The principle diagram of AC withstand voltage test is shown in Figure 2-23.

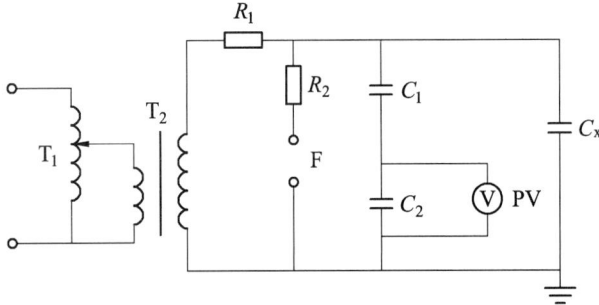

T_1–voltage regulator; T_2–test transformer; R_1–protective resistor; R_2–sphere gap protective resistor; F–spark gap; C_x–test object; C_1, C_2–capacitive voltage divider high and low arms; PV-voltmeter.

Figure 2-23　AC withstand voltage test principle diagram

In the AC withstand voltage test, measures such as connecting the protective resistor R_1, installing discharge spark gaps F, and setting overvoltage protection are taken to ensure the safety of the test personnel and other relevant staff and prevent equipment and personal accidents. The protective resistor is connected to the output terminal of the test transformer to limit short-circuit currents and suppress over voltages caused by high-frequency oscillations during discharge. The action voltage of the overvoltage protection is set at 1.1 to 1.5 times the test voltage, and the action current of the overcurrent protection is set at 1.3 to 1.5 times the current in the specimen.

Once overvoltage or overcurrent occurs, the test power supply is disconnected. The test specimen is connected to the protective spark gap G at both ends to prevent the occurrence of resonant over voltages in the test circuit, and the discharge voltage is set at 1.10 to 1.50 times U_t (the test voltage). AC withstand voltage test wiring diagram is shown in Figure 2-24.

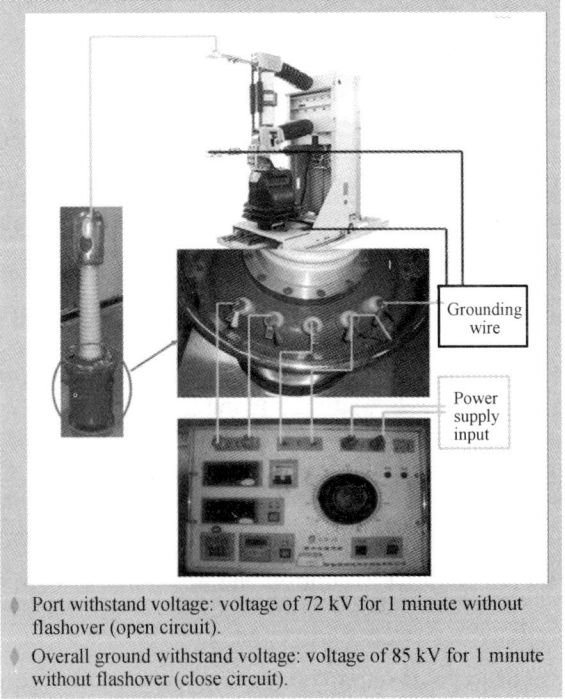

Figure 2-24　AC withstand voltage test wiring diagram

2. Test process

First, adjust the protective spark gap. Remove the high-voltage lead connected to the test object and short-circuit the ammeter connected to the grounding terminal of the test transformer. Adjust the distance of the protective spark gap and then close the test power switch. Slowly and evenly increase the voltage by adjusting the voltage regulator. The voltage should be increased from zero, and a rapid increase in voltage is allowed before reaching 40% of the test voltage. After that, the voltage should be increased evenly at a rate of 3% per second of the test voltage.

Adjust the discharge voltage to 1.1 to 1.2 times the test voltage and then decrease it to the test voltage value. Keep the voltage constant for 1 minute and observe if there are any abnormalities in various meters. Finally, decrease the voltage to zero and disconnect the test power switch.

3. Withstand Voltage Test

After completing the above steps, securely connect the high-voltage lead to the test specimen and close the power switch to initiate voltage increase. The voltage can be quickly and uniformly increased until it reaches 0.75 times the test voltage. Subsequently, continue to increase the voltage at a rate of 2% per second until the desired test voltage is reached. The test voltage should be sustained for a specified duration, usually 1 minute. Once the withstand voltage test is completed, promptly reduce the voltage to zero, open the power switch, ground the test specimen, and avoid abruptly cutting off the power without voltage reduction.

Throughout the process of voltage increase and voltage endurance, closely monitor various instrument readings for any abnormalities, as well as inspect the test object for signs of arcing, smoking, burning, charred odors, or discharge sounds. If any of these phenomena occur, immediately and uniformly decrease the voltage to zero, disconnect the power switch, ground the test specimen, and conduct analysis and assessment for potential issues.

After the withstand voltage test, thoroughly examine the test object and perform insulation resistance testing to evaluate the insulation condition post-test. For organic insulation, after the withstand voltage test, power-off, grounding, and discharge, the test personnel can safely touch it by hand to check for any signs of heating.

When the test object is made of organic insulation material, immediately touch it after the test discharge. If heat is generated, it indicates poor insulation. Address the issue and conduct the test again. For equipment with laminated insulation or organic insulation materials, if the insulation resistance after the withstand voltage test decreases by 30% compared to before the test, inspect the test sample to determine its qualification.

During the voltage increase test, if significant fluctuations in the voltmeter or ammeter pointer, an increase in current without a voltage change, smoke, burning smell, flashover, or breakdown of the test sample are observed, immediately stop the voltage increase. After reducing the voltage and cutting off the power, investigate the cause. If it is determined that these phenomena are caused by insulation failure, the AC withstand voltage test is considered unsuccessful.

During the test, if surface tracking discharge or air discharge occurs due to factors such as high humidity, temperature, or surface contamination, it should not be considered as a failure of the internal insulation of the test sample. After cleaning and drying, conduct the test again. Set up protective barriers and have someone monitor them. If any abnormality is found, immediately turn off the power, stop the test, and find out the cause.

III. Operation of Variable Frequency Resonance Withstand Voltage Test

1. Test Wiring

The wiring for the variable frequency series resonance withstand voltage test is shown in Figure 2-25.

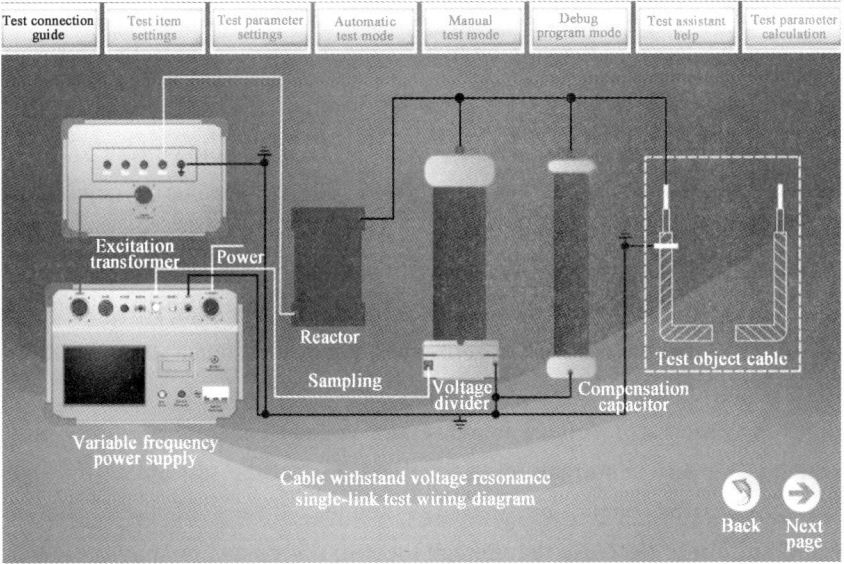

(a) Wiring for variable frequency series resonance withstand voltage test of cross-linked polyethylene cables.

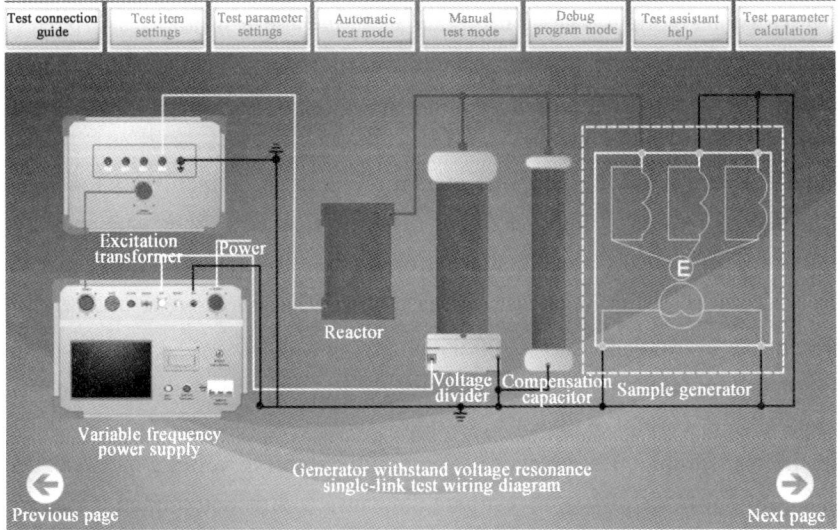

(b) Schematic diagram of the wiring for variable-frequency series resonance test of a thermal power generator

Figure 2-25　The wiring for the variable frequency series resonance withstand voltage test

2. Test procedures

(1) Arrange the test equipment properly, set up a high-voltage charged area and a test operation area, and ground the excitation transformer, resonant reactor, the enclosure of the tested equipment, and the grounding terminal of the potential divider.

(2) Before the test, measure the insulation resistance of the tested equipment. Then, wire according to Figure 2-4 and check the wiring and the gear position of the potential divider.

(3) Check that the capacity of the test power supply meets the requirements of the test. First, close the test power supply switch, then close the control power switch of the frequency converter power supply. Once stabilized, close the main circuit switch of the frequency converter power supply and set the protection voltage to 1.1–1.2 times the test voltage.

(4) Start the voltage boost. According to regulations, gradually increase the voltage uniformly from zero. Begin by rotating the voltage adjustment knob to adjust the output power ratio to 2% or 3%–5% of the test voltage. Change the system frequency by rotating the frequency adjustment knob and observe the values of excitation voltage and test voltage. When the excitation voltage is at the minimum and the test voltage is at the maximum, this frequency represents the resonant frequency of the system.

(5) Once the system is in resonance, uniformly adjust the voltage to the test voltage as required. During the voltage boosting process, closely monitor the high-voltage circuit and listen for any abnormal sounds from the tested device. After reaching the designated test duration, the detailed parameters during the test will be displayed on the test result interface. Reduce the voltage to zero, disconnect the main circuit, control circuit, and power supply switches. Open the test power switch and discharge the tested device. The test is complete.

3. Operating Instructions for Series Resonance Test Equipment

After starting up, first configure the parameters such as test frequency, voltage, and duration. The starting frequency should be set to the initial frequency for automatic tuning, with a lower limit of 20 Hz and an upper limit of 200 Hz. The ending frequency should be set to the final frequency for automatic tuning, with a lower limit of 100 Hz and an upper limit of 300 Hz. The scanning range during the test should be set from 30 Hz to 300 Hz. The starting voltage should be set to the initial value of the input voltage during tuning. For devices with low Q values, such as generators, motors, and overhead busbars, the initial value should be set between 50 V and 70 V. For devices with high Q values, such as power cables, transformers, GIS, etc., the initial value should be set between 30 V and 50 V.

According to the testing requirements, configure voltages and withstand test durations for Stage 1, Stage 2, and Stage 3.If there are no staged withstand tests, simply set the voltage value and corresponding test time for one stage, while setting the voltages and test times for other stages to 0.The "divider ratio" of the capacitive voltage divider should be set to 3,000. Overvoltage protection can be set to the limit value of the test voltage, automatically terminating the test if the voltage exceeds this value, typically set 10% higher than the test voltage. Overcurrent protection should be set to the maximum value of the low-voltage output current. If the actual test current is unknown, it is generally set to the rated current of the equipment. Flashover protection is set to the error value of the breakdown voltage.

The parameter configuration is complete, and the test can begin. Click "Tuning" to let the system automatically search for the resonance point and tune the voltage up. If there are any abnormal situations, click the "Lower Voltage" button to stop the operation. The red line represents the voltage curve, while the green line represents the frequency curve. When the voltage ($U_{resonance}$) reaches the withstand value of the test, the system will automatically start the withstand timing. When the set withstand time is reached, the system will automatically lower the voltage. The test is considered complete when the voltage ($U_{resonance}$) drops to 0.

For manual testing: After setting the "Test Parameters," click "Manual Test" to enter the "Manual Test" interface. First, click "Boost Voltage" to raise the "$U_{resonant}$" to 10 V. Then, click "Boost Frequency" to find the resonance point. After finding the resonance point, click "Boost Voltage". When the $U_{resonant}$ reaches the withstand value set, click "Withstand Voltage Time", and the system will start timing. When the "Withstand Voltage Time" is over, click "Lower Voltage Stop", and the system will automatically lower the voltage. If any emergencies occur during the testing process, click "Emergency Stop". After the emergency stop, click "Fault Reset". During manual voltage up and frequency tuning, select voltage step adjustment and frequency step adjustment according to the test situation. Click "Parameter Calculation" to calculate the parameters of inductance, capacitance, and frequency, as shown in Figure 2-26..

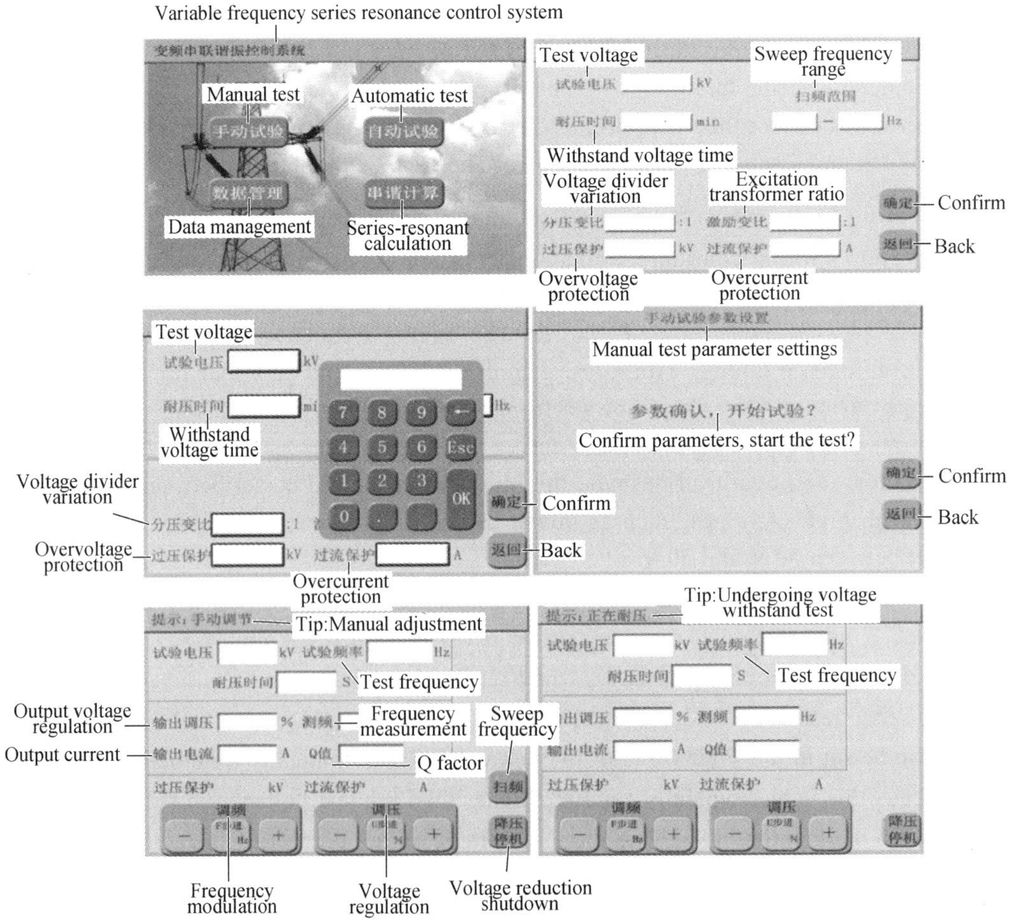

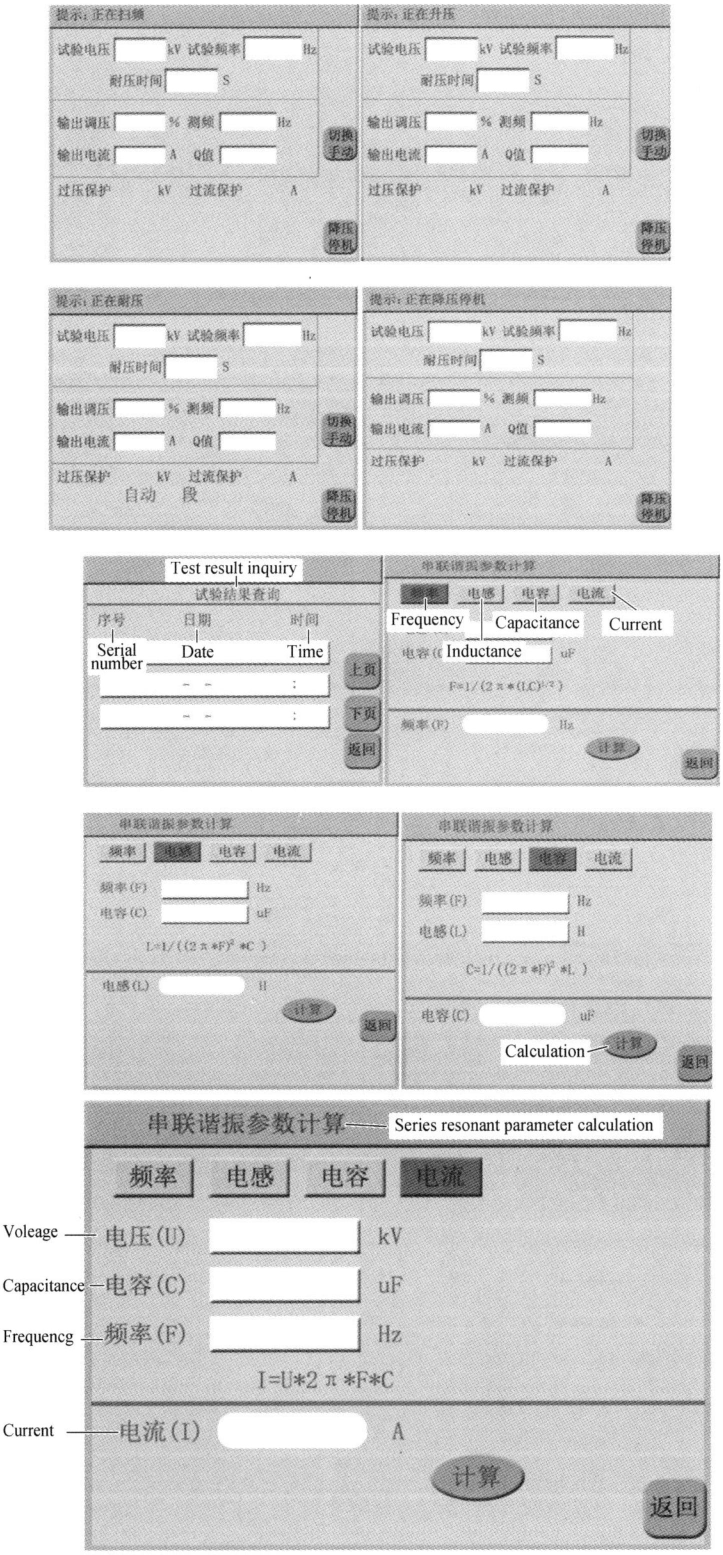

Figure 2-26 Operating steps of series resonance test device

Ⅳ. Notes on AC Withstand Voltage Test

For the newly-filled oil-immersed transformer or oil-immersed transformer after maintenance, they need to be left to stand for a certain period of time until the bubbles escape before withstanding voltage testing. For electrical transformers above 500 kV, they need to stand for more than 72 hours; for transformers above 220–330 kV, they need to stand for more than 48 hours; for other electrical transformers, they need to stand for more than 24 hours. After the AC withstand voltage test of oil-immersed transformers and other products, in order to check whether the product has been penetrated or suffered insulation damage during the test, the oil object of the oil-immersed product should be analyzed by chromatography or partial discharge measurement in a timely manner to detect any abnormal situations.

During the withstand voltage test, the voltage must start from zero and increase steadily. When the test voltage reaches 75% or above, it should be slowly increased at a rate of 2% per second. Sudden voltage increases from a non-zero position or sudden power disconnection at higher voltages should be avoided to prevent damage to the test specimen caused by transient over voltages. During the voltage increase process and the duration of the test voltage, pay attention to observing the various parts of the test instrument and test circuit.

Generally, the test voltage and current in the high-voltage circuit should increase proportionally. If there is a slight increase in voltage accompanied by a sharp increase in current or an increase in current accompanied by a decrease in voltage, it indicates that the test circuit may be in resonance. In such cases, the test voltage should be immediately reduced to zero, the power supply should be disconnected, the high-voltage outlet bushing of the test transformer should be grounded, and the impedance potential transformer and voltage regulator should be replaced or the parameters of the test transformer load should be changed before conducting the test again.

In AC withstand voltage tests, if there are problems with the specimen, test equipment, etc., the meter of the test instrument will swing significantly, the specimen may emit smoke, discharge, and a burnt odor, accompanied by discharge sounds or other abnormal noises, and the protective spark gaps will discharge, and overvoltage and overcurrent protections will activate. If any of these phenomena occurs, the test voltage should be immediately reduced to zero, the power supply should be disconnected, and a grounding wire should be installed. After identifying the cause and troubleshooting, the AC withstand voltage test can be conducted again. If these phenomena are caused by weak insulation in the cable of the specimen, the withstand voltage test is considered unqualified. If it is determined that pollution is caused by factors such as high air humidity or dirt on the surface of the specimen, the specimen cable should be cleaned and dried before conducting the test again.

Due to the AC voltage being a sine wave, the test voltage value should be taken as the peak value of the AC test voltage divided by $\sqrt{2}$ to eliminate measurement errors caused by voltage waveform distortion. During AC withstand voltage tests at power frequency, the continuous test time applied to the specimen is 1 minute, which will neither damage the equipment insulation nor expose insulation defects. In order to improve test speed during factory inspections, and within the allowable insulation level of the equipment, the duration can be shortened to 1 second when the test voltage is increased by 25%.

(1) The capacity of the test power supply must meet the test requirements.

(2) To reduce corona losses, high-voltage leads should use large-diameter metal hoses and be kept as short as possible.

(3) Overcurrent and overvoltage protection in the test apparatus must be sensitive and reliable, and a lightning arrester should be installed on the excitation transformer's high-voltage side.

(4) During testing, the resonance frequency should be adjusted at a lower voltage before increasing the voltage for the test.

(5) Humidity has a significant impact on the test, so the test should be conducted in dry weather conditions.

AC withstand voltage test is a destructive test. Before the test, insulation resistance, absorption ratio, leakage current, dielectric loss angle, and insulating oil tests must be conducted on the object. Only when the test results are normal then the AC withstand voltage test can be carried out. If insulation defects such as dampness and partial defects are found in the equipment, treatment should usually be carried out first before conducting the withstand voltage test to avoid unnecessary insulation breakdown. The specimen's insulation resistance should be measured before and after the AC withstand voltage test. The insulation resistance measured after the AC withstand voltage test should not show any significant change compared to that before the test.

In the AC withstand voltage test, the specimen is generally considered qualified if no breakdown occurs; otherwise, it is considered unqualified. The occurrence of breakdown in the object can be analyzed according to the following conditions. The relative humidity should not exceed 80% during the AC withstand voltage test, and the ambient temperature and humidity should be recorded.

During the withstand voltage process, if surface flashover or air discharge occurs due to the effects of insulation dampness or surface dirt on the object, it should not be easily considered as unqualified. Instead, it should undergo cleaning and drying treatment before conducting the withstand voltage test. If after excluding external influencing factors, surface flashover or local discharge heating still occurs during the withstand voltage test, it indicates that there are insulation problems, such as aging or excessive surface loss.

Module 4　Training

What is the essence of the AC withstand voltage test?
_____
_____
_____

Worksheet 5 Partial Discharge Test

Module 1 Operating Worksheet: Partial Discharge Test

(I) Test Name and Instrument	(II) Test Objects
Partial Discharge Detector **Handheld Ground Wave Partial Discharge Detector**	Partial discharge testing can be conducted on power equipment including the power transformer with voltage levels of 110 kV and above, power current transformer with voltage levels of 35 kV and above, GIS (Gas Insulated Switchgear), cable, bushing, lighting arrester, coupling capacitor, and other power equipment
(III) Test Purpose	(IV) Test Steps
(1) Detect whether there are gaps or impurities inside the insulation sensitively. (2) Discover defects in equipment structure and manufacturing processes, such as sharp corners and burrs on metal components. (3) Detect the poor electrical connections between internal metal grounding components or conductive parts with defects in the insulation, in order to eliminate these defects and prevent partial discharge from damaging the insulation	(1) Verify electricity, set up the grounding wire, and take necessary safety precautions. (2) Set up a guard, and have the testing personnel and safety inspectors ready for the test. (3) Detect the presence of abnormal partial discharge levels using a handheld partial discharge detector
(V) Precautions	(VI) Technical Standards
(1) Before starting the test, the testing personnel should check the wiring thoroughly to ensure accurate connections and reliable grounding of the instrument and equipment. (2) Shield the high-voltage terminals. (3) Before the test, ensure that there is no floating potential. (4) Ensure a safe distance between the high-voltage equipment and surrounding equipment. (5) After the test, restore the various connections, tap shielding ends, and covers of the test specimen	According to the rules specified by the national standard for high-current generators, the allowable value of partial discharge (Q) is determined based on the imposed line-end voltage (U_2) for the test specimen: (1) For transformers, the discharge level should not exceed 300 pC at $U_2 = 1.3U_m$ and should not exceed 500 pC at $U_2 = 1.5U_m$. (2) For solid insulation current transformers, the discharge level should not exceed 100 pC at $U_2 = 1.1U_m/\sqrt{3}$, and if necessary, it should not exceed 500 pC at $U_2 = 1.1U_m$. For oil-immersed current transformers at 110 kV and above, the discharge level should not exceed 20 pC at $U_2 = 1.1U_m/\sqrt{3}$

Continued

(Ⅶ) Result Judgment	(Ⅷ) Digital Resources
According to different equipment, the corresponding voltage is applied according to the standard, and the discharge level should not exceed the specified value	Handheld partial discharge inspection device Cable partial discharge test Partial discharge patrol inspection test

Module 2 Follow Me

Ⅰ. Causes of Partial Discharge

Electrical equipment used in the power grid, such as transformer, instrument transformer, reactor, GIS motor, cable, and capacitor, are designed to withstand specific insulation voltage levels corresponding to their operating voltages. Under normal conditions, their insulation performance can withstand the operating voltage. Due to poor insulation manufacturing processes, there may be defects such as bubbles, impurities, cracks, or insulation dampness and aging after long-term operation. Such flawed insulation is prone to exhibit periodic partial discharges when subjected to high-voltage alternating electric fields.

Although the energy released during partial discharges is small and does not immediately affect the insulation strength of the electrical equipment, it can gradually degrade the insulation performance. Eventually, this degradation can lead to insulation breakdown, significantly impacting the safe operation and reliability of the power grid. To prevent equipment failures, it is essential to conduct regular partial discharge detection on electrical equipment. This helps identify potential cable equipment risks, evaluate the overall health of cables, and mitigate the occurrence of equipment malfunctions, as shown in Figure 2-7.

Partial discharge serves as an early warning sign for impending faults in electrical equipment, such as transformers. During the initial stage, partial discharges typically have low intensity and progress slowly. Before complete insulation breakdown occurs, partial discharges can persist for several months or even years. Detecting partial discharges promptly allows for timely intervention, thus minimizing the risk of serious accidents. Figure 2-28 illustrates the effects and detection methods of partial discharge in power equipment.

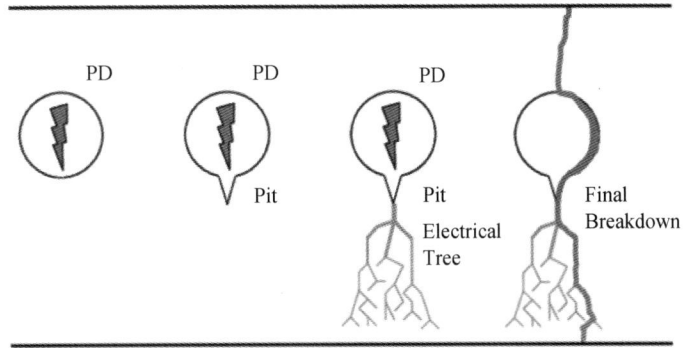

Figure 2-27　Process from Partial Discharge to Breakdown

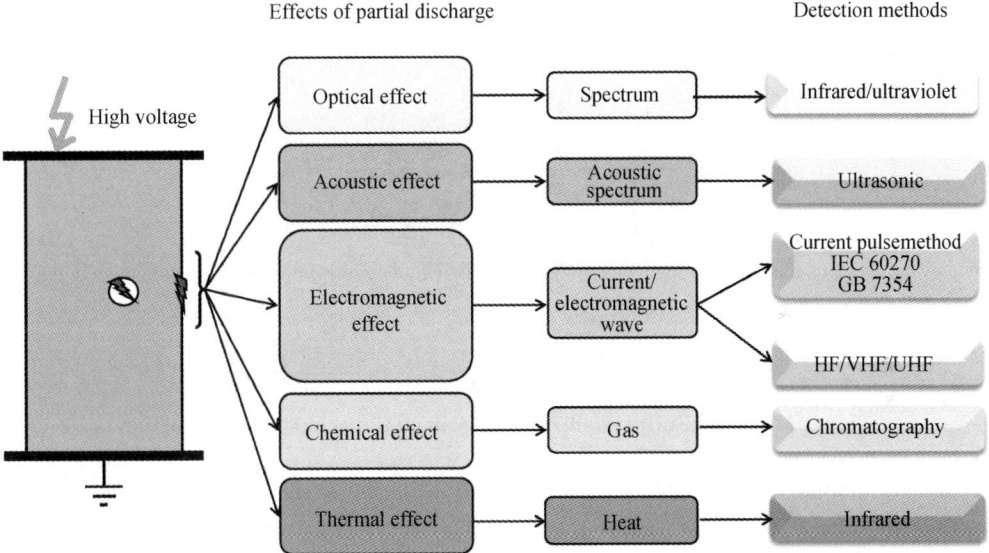

Figure 2-28　Effects and detection methods of partial discharge

Partial discharge refers to the phenomenon of non-penetrating discharge between electrodes in a local position during the operation of high-voltage electrical equipment, which does not cause complete insulation breakdown. Partial discharge includes three discharge forms, namely internal discharge (inside the dielectric), surface discharge (on the surface of the dielectric), and corona discharge (at the tip of the electrode).Long-term occurrence of partial discharge under operating voltage can eventually lead to insulation accidents in equipment due to the cumulative effects. Partial discharge is one of the maintenance methods for preventing electrical equipment failures, and therefore new or operating equipment should also be tested regularly as required. Figure 2-29, 2-30 shows GIS ultra-high frequency partial discharge live detection.

Figure 2-29 GIS

Figure 2-30 GIS ultra-high frequency partial discharge live detection

There are mainly two methods for on-site partial discharge testing: offline partial discharge withstand voltage test and online partial discharge live test. The offline partial discharge withstand voltage test is aimed at 220 kV main transformers, GIS, etc. It applies a non-partial discharge test transformer with 1.1 times the power frequency withstand voltage test. Non-partial discharge test transformers are transformers without partial discharge. Currently, there are two types of non-partial discharge transformers, one is an inflatable test transformer, and the other is an oil-immersed test transformer. It is difficult to make other test transformers into non-partial discharge transformers.

The high-voltage lead of the withstand voltage equipment should be equipped with a corrugated pipe; the outdoor terminals of the test phase and non-test phase of the withstand voltage cable should be equipped with an equalizing ring, and the non-test phase terminal should be grounded. The offline partial discharge test is used to check whether the electrical equipment structure is reasonable, the level of craftsmanship is good or bad, and internal insulation defects, including local excessive electric field strength inside the insulation, sharp corners on metal components, impurities mixed in the insulation, internal metal grounding, poor electrical connection between conductive media, and other local defects.

Transformer partial discharge is usually carried out after destructive testing, and it is conducted for faults found on-site in transformers of 220 kV and above, combined with DC resistance measurement, induced withstand voltage test, transformer characteristic test, and other comprehensive analysis to determine its insulation status, in order to discover insulation defects such as materials or manufacturing processes. Online partial discharge live testing is to install detection instruments on-site to detect possible partial discharge in electrical equipment such as transformers, high-voltage switchgear, GIS, cables, etc. under operating voltage.

II. Principle of Partial Discharge Detection

The following mainly introduces on-site online partial discharge detection. The mainstream methods for online partial discharge detection are ultra-high frequency (UHF) and acoustic emission (AE) detection technology. As partial discharge is a kind of pulse wave, it will produce a series of physical phenomena and chemical changes such as light, sound, electrical and mechanical vibrations in the interior and surrounding space of the electrical equipment. By detecting these changes, the insulation state inside the electrical equipment can be monitored. The frequency band of UHF detection is usually 300 MHz to 3,000 MHz. UHF partial discharge detection can detect suspended discharge, corona discharge, free particle discharge, internal insulation discharge in equipment, etc. During partial discharge, a large and steep pulse current is generated due to the neutralization of positive and negative charges, which excites electromagnetic waves up to several GHz. The partial discharge is detected by receiving the UHF electromagnetic waves during the partial discharge process through an antenna sensor. The two frequencies of partial discharge, namely, radio frequency interference (RFI) and acoustic emission (AE), are shown in Figure 2-31.

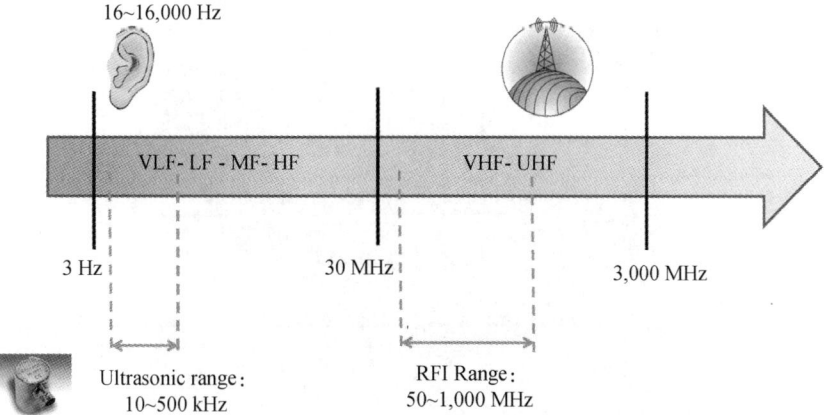

Figure 2-31 Schematic Diagram of Two Frequencies, Radio Frequency interference (RFI) and acoustic emission (AE), for partial discharge detection

The advantage of UHF detection is its high sensitivity and strong resistance to low-frequency corona interference. It can use high-bandwidth and high-sampling-rate oscilloscopes to achieve time delay positioning of discharge sources, thus achieving precise localization. However, its disadvantages are that it is susceptible to interference from UHF signals in the environment and the shielding effect of metal. During online partial discharge inspection, the detection positions are generally selected in easy-to-discharge locations inside the GIS, such as circuit breakers, lower parts of high-voltage bushings, or busbar gaps.

Since the principle of UHF detection for partial discharge detection is to measure the high-frequency current pulses generated by partial discharge in the insulation structure of the test specimen under a certain voltage, it is necessary to eliminate high-frequency pulse signals caused by environmental high temperature, on-site interference, large equipment vibration, personnel movement, mobile phones, engines, and other high-frequency signals during actual testing. Otherwise, it will decrease the detection sensitivity and easily cause misjudgment. On-site detection should be strictly carried out according to regulations and operating instructions, following the detection process, ensuring complete detection of important areas, preventing missed, insufficient, or erroneous detection. Data collection and analysis should be completed on-site, combined with reasonable analysis of the detection indicators. As shown in Figure 2-32.

The handheld partial discharge inspection instrument integrates two modes of partial discharge detection, radio frequency interference (RFI) and acoustic emission (AE), and can quickly complete insulation defect detection and localization of various electrical equipment in the entire substation, including GIS, transformer, switchgear, and cable termination. As shown in Figure 2-33.

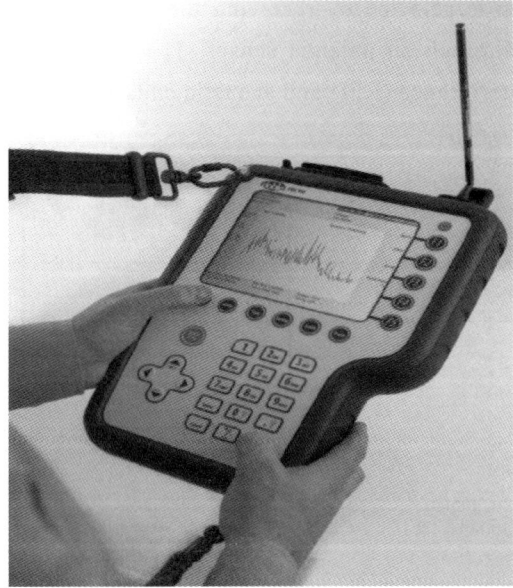

Figure 2-32　Handheld Partial Discharge Inspection Instrument

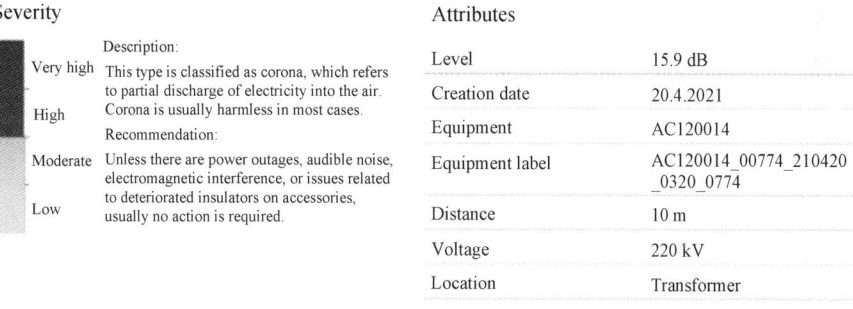

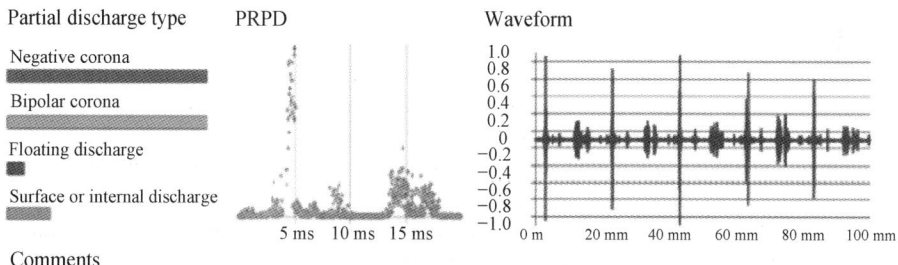

Figure 2-33　Type and severity of partial discharge in transformers identified by UHF detection

　　The principle of ultrasonic detection for partial discharge is similar to UHF detection. When partial discharge occurs inside electrical equipment such as GIS, it generates steep current pulses that cause instantaneous expansion of SF_6, resulting in an explosive effect. After the discharge, SF_6 returns to its original volume. The volume changes caused by the expansion and contraction of the insulation during partial discharge create ultrasonic waves. Ultrasonic waves have a frequency above 20 kHz. They propagate in the form of spherical waves with the partial discharge point (source) as the center. By using sensors within the range of 20 kHz to 100 kHz attached to the metal casing of GIS, the ultrasound signals are detected, enabling the localization of partial discharge and detection of vibration-related defects (mechanical vibration and hysteresis vibration) as well as partial discharge-related defects (tip discharge, free metal particle discharge, suspended potential discharge).

The UHF RFI mode detection process consists of four steps: frequency scanning for frequency domain analysis, time domain analysis to determine the presence of partial discharge, confirmation of partial discharge type, and localization of partial discharge source through frequency domain analysis or power meter.

The ultrasonic mode detection process consists of three steps: detecting partial discharge activity through bar graph mode, distinguishing partial discharge types through phase mode, and identifying free metal particle discharge phenomena through pulse mode.

Module 3 Workshop

I. Partial Discharge Operation Process (see Table 2-9)

Table 2-9 Operational steps for partial discharge testing

ID	Testing Stage	Operational Steps (using transformer wiring as example)
1	Site preparation	(1) Consult the test data of the test object. (2) Inform safety and technical measures. (3) Inform personnel organization and division. (4) Familiar with construction site (5) Check the implementation of safety measures
2	Preparation before testing	(1) Prepare and check equipment and instruments. (2) Prepare and check safety supplies. (3) Perform routine inspection of specimen. (4) Insulate the primary and secondary windings of the transformer, short-circuit the secondary winding, and ground it. (5) Test insulation of transformer tap shielding end. (6) Check reliable grounding of transformer tap shielding end
3	Conducting the test	(1) Ensure the absolute safe distance between voltage-adding equipment, transformer, and surrounding equipment. (2) Ensure the accuracy of test wiring. (3) Check that the grounding wire of the test instrument and equipment is absolutely reliable. (4) The test operation procedure should be in accordance with regulations. (5) Record the test data truthfully and accurately
4	Test completion	(1) Remove the test wiring. (2) Recheck the insulation of the primary and secondary windings of the transformer. (3) Recheck the insulation of the transformer tap shielding end and restore the tap grounding end. (4) Fill out the test record

Partial discharge tester is shown in Figure 2-34.

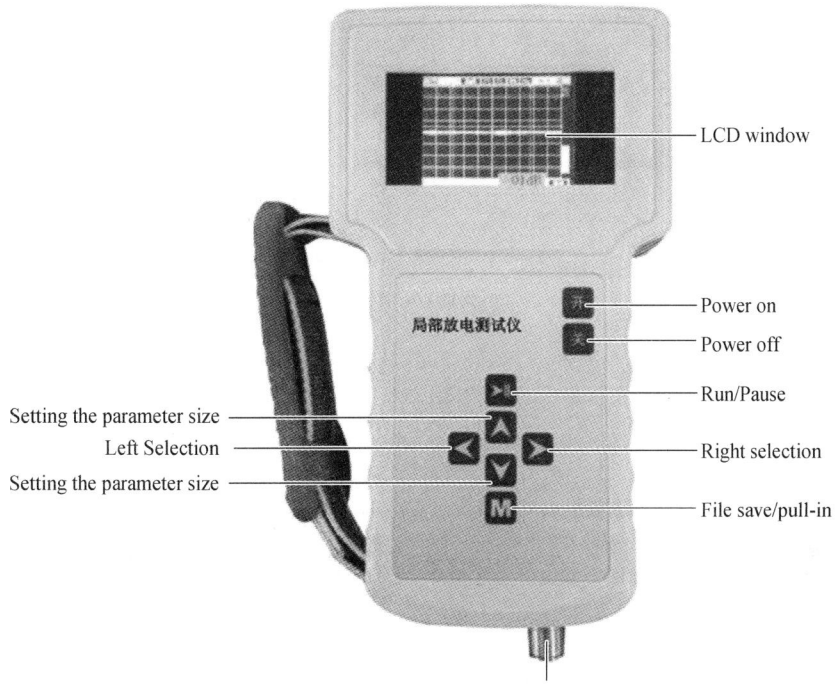

(a) Front view of main unit

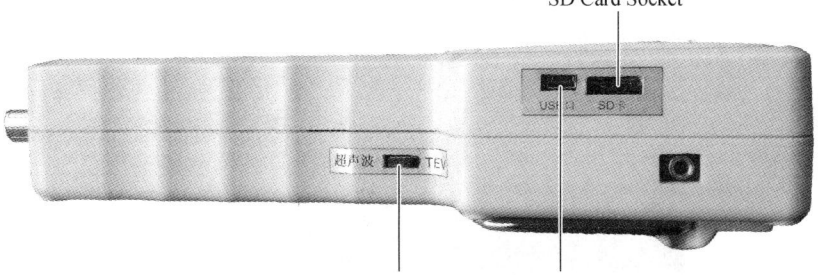

(b) Side view of main unit

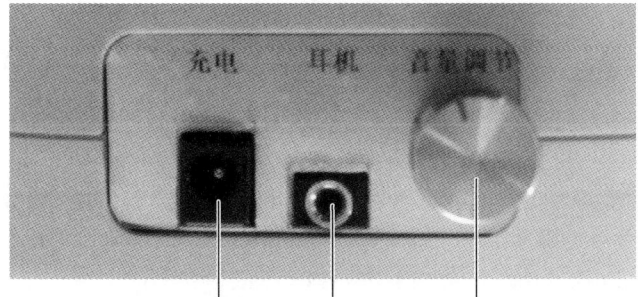

(c) Bottom view of mainframe

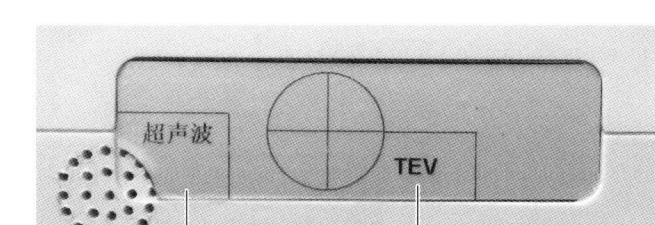

(d) Top view of mainframe

Figure 2-34 Partial discharge tester

Notes:
① TEV Test Area: In TEV test mode, align this area to the test point.
② Ultrasonic test area: In the ultrasonic test mode, align this area to the test point.

Ⅱ. Process of Partial Discharge

Turn on the instrument and ensure that the TEV sensor is away from the metal electrode to avoid affecting the self-test function. Select the TEV mode. While measuring, ensure that the TEV probe is in contact with the metal electrode to be measured and kept vertical, as shown in Figure 2-35.

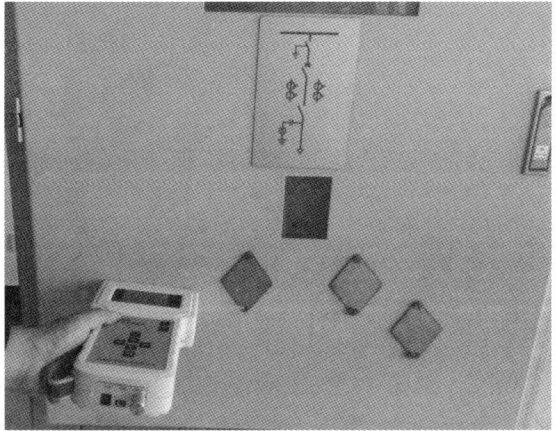

Figure 2-35 Process of partial discharge

The measurement focus is on the central positions of cable boxes, current transformer chambers, busbar chambers, circuit breakers, and potential transformers in each panel of the switchgear. During the measurement, record the first set of readings at each position. If the measured amplitude is 10 dB higher than the background interference level and the amplitude itself is greater than 20 dB, three consecutive sets of readings should be recorded.

When conducting ultrasound measurements, first turn on the instrument and select the ultrasound mode from the menu. Insert the provided headphones and adjust the volume. The readings will continuously update on the display screen. Start by measuring the background noise and record it. If the readings become too high, reduce the gain. To inspect the switchgear, point the ultrasound sensor towards the switchgear, especially at the ports of circuit breakers, inflatable cable boxes, potential transformers, and air gaps in busbar chambers. Discharges can be identified by the hissing sound emitted through the headphones, similar to the sound of frying pans. The ultrasound device is equipped with a laser aiming function, allowing detection of surface discharges at both close and distant ranges. The reflector is transparent and helps precisely aim at the target for measurement.

The ultrasound mode display interface is shown in the Figure 2-36.

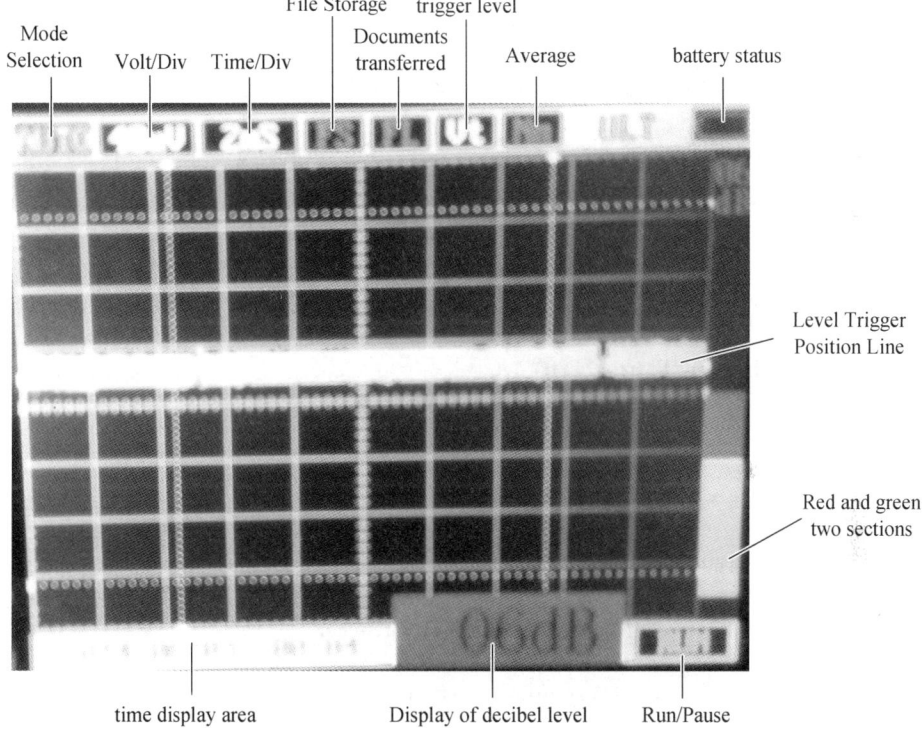

Figure 2-36　Ultrasonic mode display interface

The ultrasound interface displays two color segments in green and red. A measurement value of <6 dB is displayed in green, while a measurement value of ≥6 dB is displayed in red. The default value for the red segment is 6 dB, with a setting range between 3 and 10 dB.

The TEV mode display interface is shown in Figure 2-37.

The TEV mode displays three color segments in red, green, and yellow. The default value for the red segment is 29 dB (with a setting range of 25–34 dB), while the default value for the yellow segment is 20 dB (with a setting range of 16–24 dB). A measurement value of <20 dB is displayed in green, a measurement value of ≥20 dB and <29 dB is displayed in yellow, and a measurement value of ≥29 dB is displayed in red.

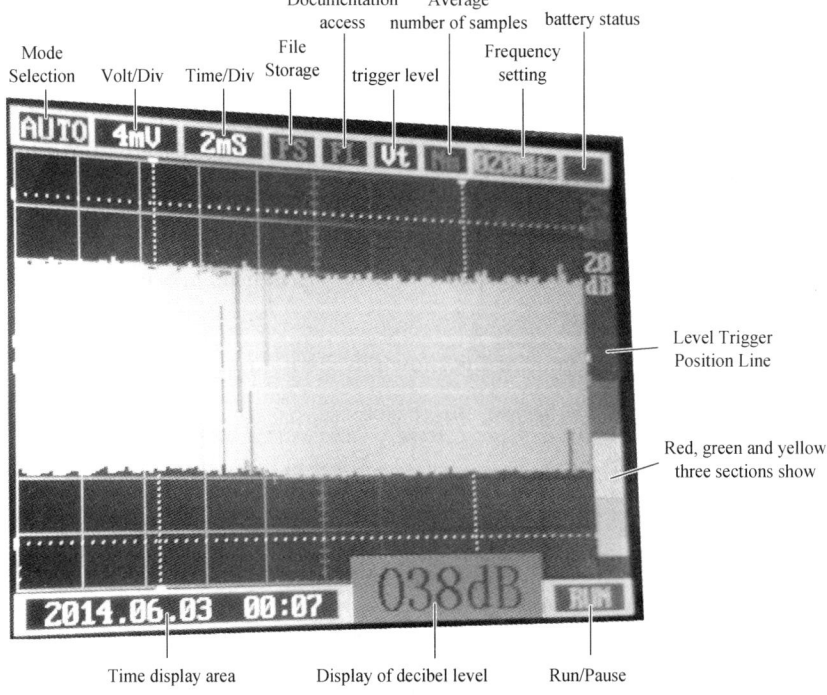

Figure 2-37　TEV mode display interface

Ⅲ. Data analysis (see Table 2-10)

Table 2-10　Data analysis

TEV Reading	Conclusion
(1) High background readings, greater than 20 dB	(1) High levels of noise may mask discharges within the switchgear. (2) If it is suspected that external influences are causing this, external sources of interference should be eliminated before retesting, or a partial discharge monitoring device should be used to identify any discharges within the switchgear
(2) all readings from the switchgear and background reference are below 20 dB	No significant discharges. Recheck once a year
(3) The reading from the switchgear is 10 dB higher than the background level, and the absolute value of the reading is above 20 dB (not 20 dB higher than the background)	There is a high probability of internal discharge activity within the switchgear. It is recommended to perform further inspection using a partial discharge locator or a partial discharge monitoring device

Module 4　Traning

What insulation problems can be detected by the partial discharge test? What is its principle?

_____
_____
_____

Evaluation

ID	Test 1	Test 2	Test 3	Test 4	Test 5
Project Name	Insulation Resistance Test	DC Leakage Current and Withstand Voltage Test	Dielectric Loss Measurement	Withstand Voltage Test	Partial Discharge Test
Test instrument	Manual or electronic megohmmeter	DC high voltage generator	Anti-interference dielectric loss automatic tester	One set of AC withstand voltage test devices or one set of series resonant withstand voltage test device	One set of partial discharge measurement instrument
Test content	(1) Measure the insulation resistance of the transformer core and clamps. (2) Measure the insulation resistance, absorption ratio, and polarization index of the transformer windings. (3) Measure the main insulation and tap shielding end-to-ground insulation resistance of the transformer bushing	(1) DC withstand voltage test. (2) DC leakage current	Transformer dielectric loss measurement	AC withstand voltage test (series resonant withstand voltage test)	Transformer partial discharge test
Project requirements	(1) Explain the principle of insulation testing for oil-immersed transformers. (2) Demonstrate and explain the on-site operation process required for the test. (3) Pay attention to personal and equipment safety, ensuring that the operation process complies with safety regulations. (4) Clean up the site after the test is completed. (5) Write the test report	(1) Explain the principles of DC leakage current and DC withstand voltage testing. (2) Demonstrate on-site operation and explain the insulation structures and materials that need to be tested. (3) Pay attention to personal and equipment safety, ensuring that the operation process complies with safety regulations. (4) Clean the test specimen. (5) Write the test report	(1) Explain the principle of dielectric loss testing. (2) Demonstrate on-site operation and explain the insulation structures and materials that need to be tested. (3) Pay attention to personal and equipment safety, ensuring that the operation process complies with safety regulations. (4) Clean the test specimen. (5) Write the test report	(1) Explain the principle of AC withstand voltage testing. (2) Demonstrate on-site operation and explain the insulation structures and materials that need to be tested. (3) Pay attention to personal and equipment safety, ensuring that the operation process complies with safety regulations. (4) Clean the test specimen. (5) Write the test report	(1) Explain the principle of partial discharge testing. (2) Demonstrate on-site operation and explain the insulation structures and materials that need to be tested. (3) Pay attention to personal and equipment safety, ensuring that the operation process complies with safety regulations. (4) Clean the test specimen. (5) Write the test report
Step 1: Tool preparation	(1) Safety tools: voltage tester, insulating pole, grounding wire, connecting wires, discharge rod, warning fence, and warning signs. (2) Insulating tools: insulating gloves, insulating boots, safety helmet, safety belt, insulating mats, insulating ladder. (3) Tool inspection and arrangement: ① Divide the testing area and operation area, correctly place the testing equipment, and maintain a safe distance from the test object. ② Prepare relevant test cables, alligator clips, grounding wires, etc. ③ Prepare other tools: a multimeter, thermometer, humidity meter, etc.					

Continued

Step 2: Risk Control	(1) Before the test, ensure proper "two-wearing and three-carrying" (wear working uniform, wear insulating boots, wear a safety helmet, wear insulating gloves, and use a voltage tester). (2) Set up barriers at the testing site and hang signs outside saying "No entry, high voltage danger." (3) Take necessary precautions for working at heights if required. (4) During the test, contact with the equipment is prohibited. (5) Prior to the test, the test personnel must follow the requirements of the "Work Permit" and take necessary safety measures
Step 3: Test wiring	(1) Connect according to the requirements of the test project, and all test connections must be reliable and secure. (2) The testing instruments must be reliably grounded. (3) Sufficient safety testing distance should be maintained between personnel and equipment. (4) Establish and implement safety supervision and protection systems before and after the test
Step 4: Test operation	(1) Electrical testing must be conducted by two or more staff, with one person operating and another person supervising. (2) Enter the instrument's operation interface to set the test parameters. (3) Voltage should be slowly increased from zero during the test, and the voltage ramp-up speed should not be too fast to prevent sudden voltage increase. After the test is completed, the voltage should be lowered to zero. (4) A call-and-response system must be implemented during high-voltage testing. (5) During the step-up test, if the following abnormal conditions are observed, the test should be stopped to investigate the cause: ① Significant fluctuations in voltage or current meter readings. ② Unusual noises coming from the test specimen. ③ Evidence of burning or smoking from the test specimen. In such cases, the voltage should be immediately lowered, the power supply should be disconnected, and the test should be stopped to investigate the cause. (6) After changing the wiring or completing the test, the voltage should be lowered, power should be disconnected, and a grounding discharge should be performed on the test equipment. Only after confirming that the test specimen is reliably grounded can disassembly and reassembly of the test lines take place in the test area
Step 5: Data recording	(1) Record the nameplate and model of the recording instrument equipment. (2) Record the nameplate and model of the test sample. (3) Record the test data. (4) Record the historical test data of the test sample
Step 6: Results analysis	Organize the site and confirm that it is clean after completion. (1) The comparison of the test data and the factory test report should comply with the test procedures. (2) The comparison of the test data and the handover test report should comply with the test procedures. (3) The test data should comply with the standards of the electric power industry or national test regulations

Project 3

High Voltage Characteristic Test

Worksheet 1 DC Resistance Test

Module 1 Operating Worksheet: DC Resistance Test

(Ⅰ) Test Name and Instrument	(Ⅱ) Test Objects
DC Resistance Test **DC Resistance Tester**	Transformer, instrument transformer, generator, motor, shunt, conductor, cable, etc.
(Ⅲ) Test Purpose	(Ⅳ) Measurement Steps
(1) Effectively detect manufacturing defects such as material selection, welding, loose connection parts, lack of stock, open circuit, etc. In transformer coils. (2) Check for short circuits between layers and turns. (3) Check if the contact in each position of the tap switch is good. (4) Check the correctness of the parallel branch and whether there is any breakage. (5) Check for breakage or poor contact in the winding or lead wires	(1) Wiring: Connect the tested equipment to the local test column using dedicated test wires and connect the relevant grounding wire reliably. (2) Range Selection: Turn on the power switch and select the corresponding range on the display screen. (3) Testing: After selecting the range, press the confirm button to start the test. After a few seconds of "charging", the display screen will show "Testing", indicating that it has been charged and entered the measurement state. After a few seconds, the display screen will simultaneously show the selected range value and the measured resistance value. (4) Recording Measurement Values: Once the measurement value stabilizes, record the resistance value. (5) Ending: After the test, when the reset button is pressed, the power to the windings will be disconnected, and a discharge will occur with an audible alarm. Then, the display screen will return to the initial state. After the discharge sound, reconnect for the next measurement or remove the test wires and power cords to end the measurement

Continued

(V) Precautions	(VI) Technical Standards
(1) When testing the DC resistance of a power transformer, the metal casing of the tester must be reliably grounded. (2) Before testing an no-load tap switch, it must be reset. After the test item is fully discharged, disconnection can only be carried out to ensure personal and equipment safety and data accuracy. (3) It is strictly forbidden to disconnect the test wires during the measurement process. Wait for the instrument to reset and discharge before disconnecting to prevent harm to people and equipment. (4) Choose the appropriate range for testing and do not use over-range or under-range. If over-ranged, because the current cannot reach the preset value, the instrument will always be in the "discharging" state. If under-ranged, it will display "current too small". (5) When connecting the test clip to the lead-out terminals of the transformer winding, please note that the lead-out terminals are exposed to the air for a long time and covered with an oxide film on the surface, which may cause inaccurate measurement results. Clean the oxide film when wiring and ensure that the test clips and lead-out terminals are well connected	(1) For the transformer with a capacity of 1.6 MVA and above, the phase resistance imbalance should not exceed 2%. For windings without neutral point leads, the line resistance imbalance should not exceed 1%. (2) For the transformer with a capacity below 1.6 MVA, the phase resistance imbalance should not exceed 4%. For windings without neutral point leads, the line resistance imbalance should not exceed 2%. (3) The difference between the measured values and the previous values (measured at the time of factory or handover acceptance) for the same parts should not exceed 2%. (4) The above judgment results should be compared at the same temperature, and the influence of lead wires should be corrected. The formula $R_2 = R_1 \times (T+t_2)/(T+t_1)$ can be used to convert resistance values at different temperatures to resistance values at the same temperature. In the formula, R_1 and R_2 are the resistance values at temperatures t_1 and t_2 respectively. T is a constant, which is 225 when the conductor is aluminum wire and 235 when the conductor is copper wire
(VII) Result Judgment	(VIII) Digital resources
(1) When measuring the DC resistance of a transformer, due to the large inductance, it is necessary to charge it fully to minimize self-inductance. Wait for the DC resistance tester pointer to stabilize before reading the resistance value. (2) The result judgment of the DC resistance tester should involve both horizontal and vertical comparisons, analyzing factors such as temperature, humidity, measurement instruments, measurement methods, inter-phase differences, wiring methods, tap switches, poor contacts, and open circuits. (3) The DC resistance is directly proportional to temperature. The DC resistance measured at different oil temperatures should be converted to the same temperature for comparison. (4) The result judgment of the DC resistance tester should consider relevant factors and criteria and analyze the development and variation process of equipment measurement data. (5) The result judgment of the DC resistance tester emphasizes the analysis, judgment, and verification of comprehensive methods	 **DC Resistance of GIS On-off Coil**

Module 2　Follow Me

DC resistance test is a necessary measurement item during the production of transformer semi-finished products, factory tests, commissioning test, major repairs, and tap switch changes. Figure 3-1 is showing the transformer with neutral point tap damaged by lightning.

Figure 3-1　Transformer with neutral point tap damaged by lightning

Ⅰ. Significance of Measuring DC Resistance

After the installation of a transformer, its internal structure is sealed with insulation materials. By measuring the DC resistance of the transformer's three phases, the following defects can be detected:

① Check the welding quality of internal winding wires and lead wires.

② Check the contact quality of the tap switch at various positions.

③ Check for any breakage in the windings or lead-out wires.

④ Check the correctness of parallel branches and measure if there are any broken connections in multiple parallel-wound windings.

⑤ Check for short circuits between layers or turns.

Ⅱ. Test Standards

The test standards are based on the "Standard for Electrical Equipment Handover Acceptance Test in Electrical Installation Engineering" (GB 50150–2016). The DC resistance or loop resistance of power equipment is shown in Table 3-1 below.

Table 3-1 Qualified standards for DC resistance or loop resistance

ID	Power Equipment	Qualified Standards	Remarks
1	Transformer	(1) For the transformer above 1.6 MVA, the difference in resistance between phases should not be greater than 2% of the average value of the three phases. For windings without neutral point leads, the difference between line resistances should not be greater than 1% of the average value of the three phases. (2) For the transformer of 1.6 MVA or less, the difference between phases is generally not greater than 4% of the average value of the three phases, and the difference between line resistances is generally not greater than 2% of the average value of the three phases. (3) The change in measured value compared to the same position measured previously should not be greater than 2%	The measurement frequency for the transformer below 110 kV is every 6 years, while for the transformer of 220 kV and 500 kV, it is every 3 years. Measurements are required after major repairs involving tap switch changes under load or after maintenance of on-load tap changers
2	Instrument transformer	The DC resistance should have no significant difference compared to the initial or factory value, within ±3%	
3	Circuit breaker, disconnecting switch	The measured value of the circuit breaker should not exceed 120% of the factory value. The measured value of the disconnecting switch should not exceed 150% of the factory value	The measurement is conducted using the method of reducing DC voltage, with a current of not less than 100 A
4	Reactor	(1) The difference between the windings of the three phases should not exceed 4% of the average value of the three phases. (2) The difference from the previous measurement should not exceed 2%	
5	Generator	Stator Windings: The DC resistance values of each phase or branch of the steam turbine generator should not differ from the factory value by more than 1.5% (for the hydro turbine generator, it should be 1%). The difference in the DC resistance of the rotor windings compared to the initial measurement (during commissioning or major overhaul) is generally not more than 2%	

Module 3 Workshop

Ⅰ. DC Resistance Test of Transformer

(Ⅰ) Wiring

For single-phase measurement, take the measurement of the transformer AB phase as an example. The thick red wire is the positive electrode of the current terminal and is connected to I+. The thin red wire is the positive electrode of the voltage terminal and is connected to V+, and the test clamp corresponding to the red line is connected to the transformer A phase. The thick black wire is the negative electrode of the current terminal and is connected to I−. The thin black wire is the negative electrode of the voltage terminal and is connected to V –, and the test clamp corresponding to the black line is connected to the transformer B phase., as shown in Figure 3-2. After measuring the AB phase, measure the BC phase, CA phase, ab phase, bc phase, and ca phase in turn. If the transformer has a tap switch, the DC resistance should be measured at each position of the tap switch.

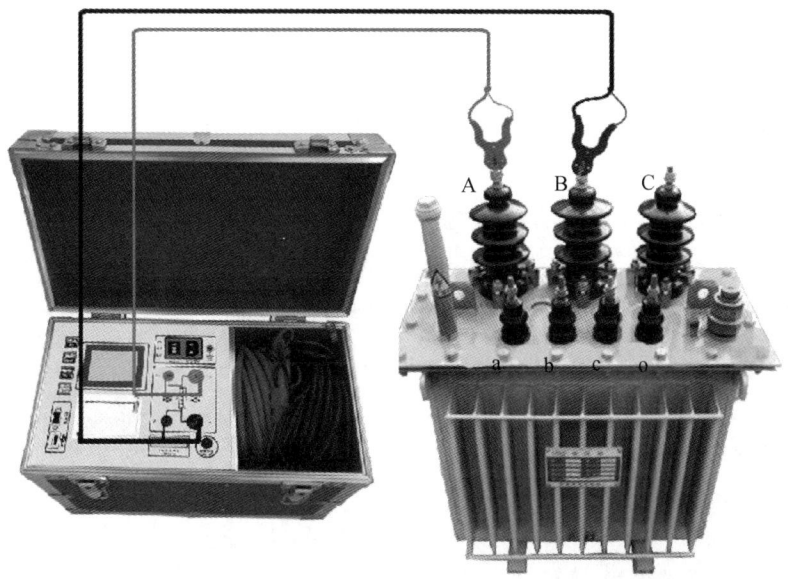

Figure 3-2　Transformer DC resistance testing wiring diagram of the high-voltage side (A-B phase)

(Ⅱ) Operation Steps for DC Resistance Testing

After powering on, enter the main interface for parameter settings. Move the cursor to select settings, query data, and select current for function selection. Click the "Select Current" button to choose different selectable test current values range (Auto, 10 A, 5 A, 1 A, 200 mA, 10 mA, <5 mA), as shown in Figure 3-3.

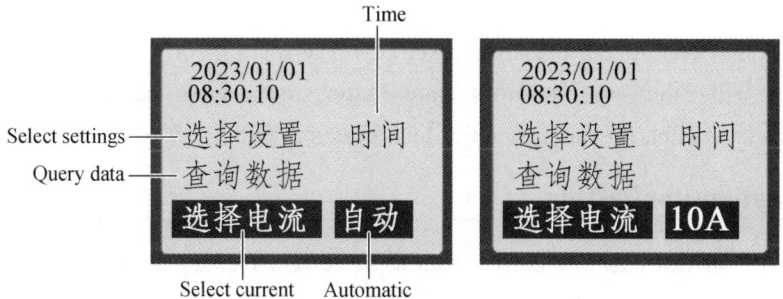

Figure 3-3　Main interface and range setting of DC resistance tester

After determining the test current based on the capacity of the transformer, press the Confirm key to start the measurement. When auto-test is selected, the instrument will automatically choose the appropriate current for testing based on the sample. After starting the measurement, the screen will display the charging and testing process, and then display the measured resistance value. Record the DC resistance measurement value after the value stabilizes. The larger the capacity of the transformer, the longer the charging time and the stabilization time, which may take between 5 and 30 minutes. The measurement process is shown in Figure 3-4.

In the figure, R is the actual measured resistance, and R_t is the converted resistance. After the test result is stable, the instrument automatically saves the test data. For the converted temperature, please refer to the temperature setting. If no temperature setting is made, it defaults to the last setting. After displaying the test result, press and hold the select key to print the test result.

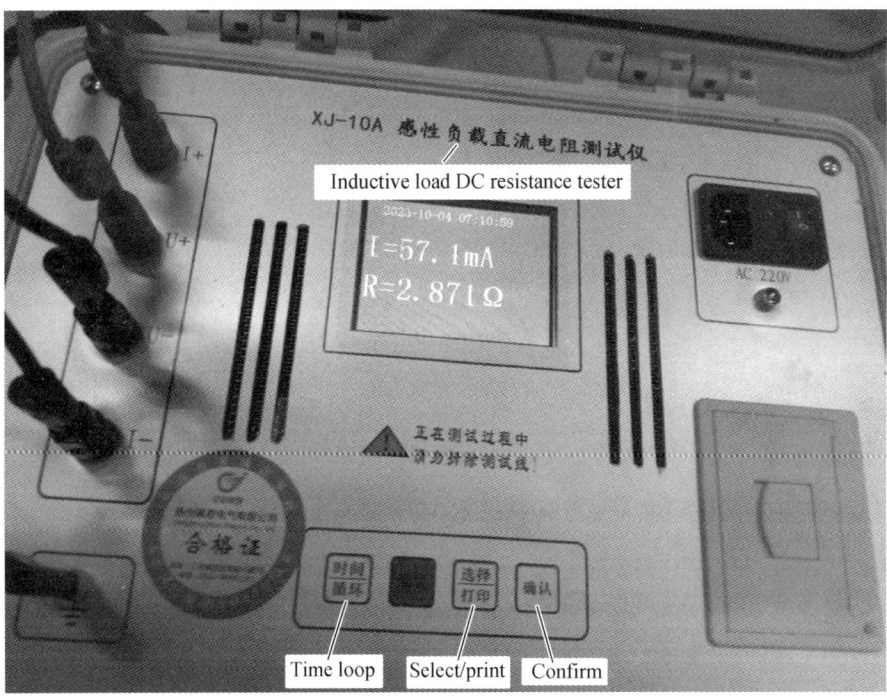

Figure 3-4   DC resistance test process and value

After the test is completed, press the reset key. The instrument will stop outputting current, and the screen will return to its initial state. After discharging, change the wiring method according to the tap switch for the next set of test measurements.

II. Precautions for Testing

Due to different test currents corresponding to certain resistance measurement ranges, it is necessary to select an appropriate test current during measurement. Excessive current may cause the winding to heat up, leading to increased error in resistance values. It may also result in significant residual magnetism in the transformer core. On the other hand, too small a test current may prolong the measurement time. Therefore, when selecting the test current, it is necessary to consider the corresponding resistance measurement range, the duration of measurement, and the potential impact of residual magnetism on the transformer.

It is prohibited to disconnect the current or voltage test leads during the measurement process. After the test is complete, press the reset key first. Only after the instrument and the transformer are discharged can the power be disconnected and the connections be removed. Otherwise, there is a possibility of electric shock to personnel.

Three-channel direct resistance measurement can simultaneously measure the DC resistance of three-phase windings. For the transformer with on-load tap changer on the high-voltage side, the DC resistance measurement can greatly shorten the testing time. However, since the three-channel measurement does not include the neutral point lead resistance, it lacks certainty in judging the condition of the neutral point lead contact in the transformer. Before the three-channel measurement, it is advisable to use single-channel measurement to measure one or two tap positions in order to compare the single/three-channel measurement values and check the condition of the neutral lead. It should be noted in the test report that a three-channel test was performed for the purpose of comparison and reference.

Therefore, the three-channel measurement method is not suitable for use in the transformer manufacturing process. The three-channel measurement takes longer for the three-phase resistance test of a single-position YN winding compared to the traditional four-terminal method and does not have the advantage of shortening the testing time. For the transformer with on-load tap changers on the high-voltage side, it is recommended to use the traditional four-terminal method for testing.

For DC resistance measurement of traction transformer windings in railway applications, the three-phase VV connection transformer (rated voltage: 110/27.5 kV, connection method: Vv0/Vv6) for conventional-speed railway or the single-phase transformer (rated voltage: $220/2 \times 27.5$ kV) for high-speed railway may look like a three-phase transformer from an external view, but it is essentially composed of two single-phase transformers. Therefore, when measuring the DC resistance, the measurement should be performed according to the steps for a single-phase transformer.

For the conventional-speed railway Vv connection transformer, the DC resistance values of the AB and BC on the primary side and T1-x1, T2-x2 on the secondary side should be measured separately. For the high-speed railway single-phase transformer, the DC resistance values of the A-X side on the primary side and a1-x1, a2-x2 on the secondary side should be measured separately. When using a three-channel direct resistance tester, the three-channel test cannot be used, and single-channel measurements should be performed on the AB and ab ends, as shown in Figure 3-5.

(a) Vv-connected three-phase railway traction transformer

(b) 220 kV combined three-phase Vv-connected traction transformer

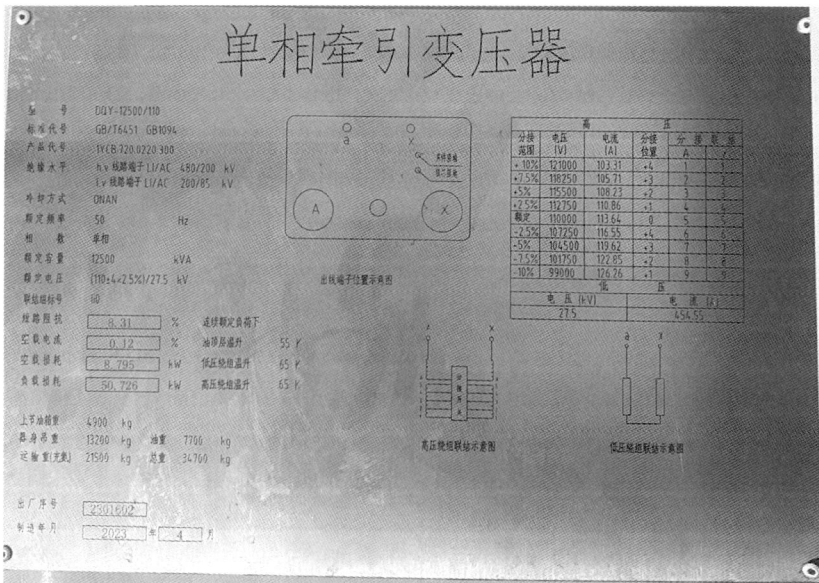

(c) Nameplate of 110 kV Single-phase transformer for conventional-speed railway

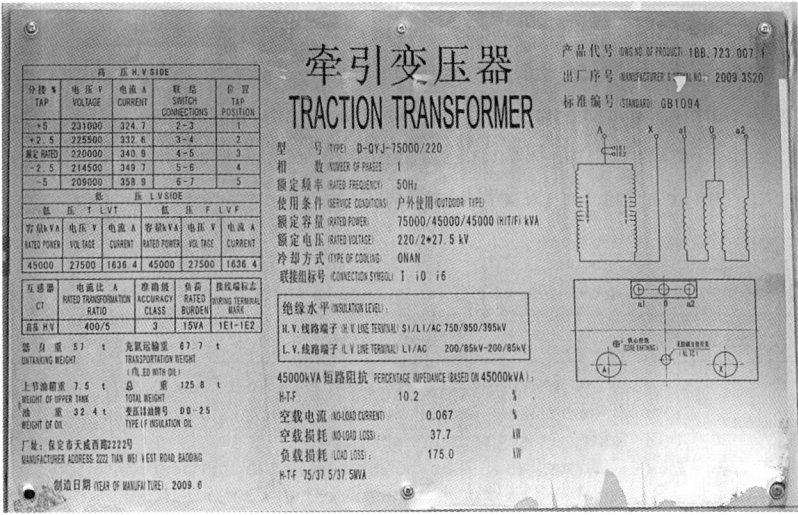

(d) Nameplate of 220 kV Single-phase transformer for high-speed railway

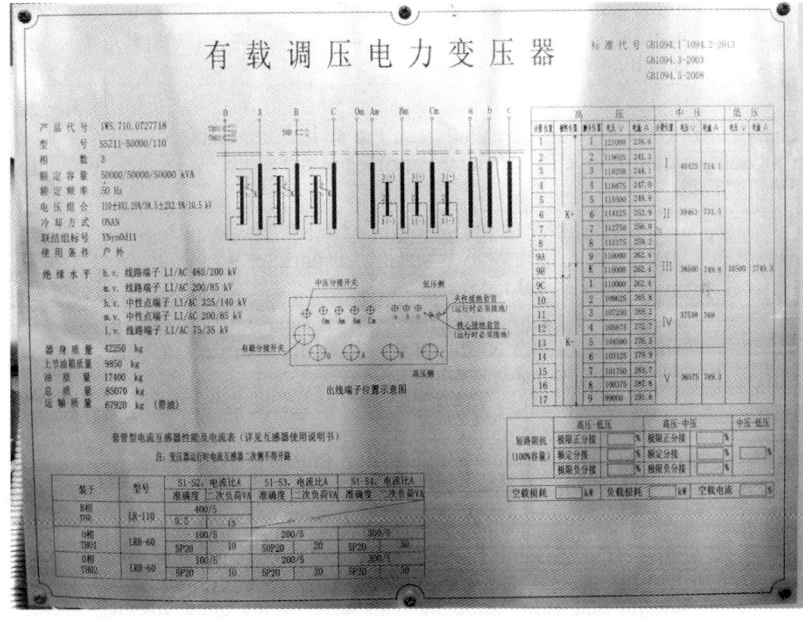

(e) Nameplate of 110 kV on-load tap changer transformer

Figure 3-5 DC resistan measurement of traction transformer windings in railway appications

Ⅲ. Judgment of Test Results

For the transformer with on-load tap changer, the DC resistance should be measured at all tapping positions. For the transformer with No-load tap changer after major overhaul, the DC resistance should be measured at all tapping positions of each winding on both sides. Preventive tests are usually conducted only in the operating range. If tapping position is changed during operation, the DC resistance must be measured again.

Some factors that can affect the measurement of transformer DC resistance include: the temperature of the upper oil layer and winding temperature directly impact the measured value of the transformer winding DC resistance, the non-test winding should be open-circuited during measurement, poor contact of tap switches, improper contact position of line clamps during measurement, etc. The reasons for the imbalance of DC resistance in the three-phase windings of the transformer are shown in Table 3-2.

Table 3-2　Reasons for imbalance of DC resistance in three-phase windings of transformer

ID	Fault Cause	Explanation	Data Performance
1	Inter-turn short Circuit	(1) YN winding: The resistance value of the faulty phase decreases, while the resistance value of the non-faulty phases remains normal. (2) Y winding: The resistance value between the phases related to the fault decreases, while the resistance value between the non-faulty phases remains normal. (3) △ winding: The resistance values between the three sets of phases all decrease, with two sets having similar resistance values and the third set having the smallest resistance value	Resistance is lower than normal
2	Poor Contact of Tap Switch	Mainly caused by unclean tap switch contacts, loss of plating layer, insufficient spring pressure, etc.	Resulting in significant deviation in three-phase winding data and a high degree of imbalance
3	Poor Welding	The welding at the junction of leads and windings is not good, or one or two strands of parallel windings are not properly welded	Resistance is higher than normal
4	Winding Open Circuit	(1) YN winding: The faulty phase cannot be charged and measured, while the resistance value of the non-faulty phases remains normal. (2) Y winding: The inter phase resistance related to the faulty phase cannot be measured and charged, while the resistance value between non-faulty phases remains normal. (3) △ winding: The resistance values between the three sets of phases all increase. The resistance of the two phases without open circuit is 1.5 times higher than the normal value, and the resistance of the open circuit phase is 3 times higher than the normal value (Notes: This refers to an open circuit within the winding, not a lead wire open circuit)	The data of the open circuit winding will show higher than normal resistance, or there may be an inability to display any data at all
5	Incorrect Wiring	Improper connection of measurement leads to transformer terminals, specifically where the voltage leads are placed on the outer side of the current leads or in the same position as the current leads during measurement. This results in the inclusion of contact resistance in the measurement value	This leads to increased contact resistance
6	Improper Measurement Method	Insufficient charging time for the winding during the measurement process, failure to disconnect other windings from the grounding electrode when measuring a specific winding, or short-circuiting of other windings	Inaccurate readings

Module 4 Training

1. What is the fundamental difference between insulation resistance and DC resistance? What is the role of DC resistance testing in high-voltage tests?

2. In addition to the transformer, which devices can be used to test the DC resistance?

Worksheet 2 Loop Resistance Test

Module 1 Operating Worksheet: Loop Resistance Test

(I) Test Name and instrument	(II) Test Objects
Loop Resistance Measurement **Loop Resistance Tester**	Various high and low switching equipment, conductor, cable, etc.
(III) Test Purpose	(IV) Measurement Steps
Measure the contact resistance of the conducting circuit in the closed state, check whether there is an oxide layer on the contact surface, and inspect whether there are any contact defects in the circuit and whether the contact is good	(1) Connect the instrument in accordance with the measurement wiring diagram requirements; (2) Turn on the power supply and require the test current to be ⩾100 A. Press the test button to start the measurement; (3) After the measurement display value stabilizes, save the data; (4) After the measurement is completed, turn off the instrument power supply after waiting for the instrument to discharge completely; (5) Hang up the discharge rod, and remove the high-voltage test connection; (6) Remove the instrument end voltage and current lines; (7) Finally, remove the grounding wire
(V) Precautions	(VI) Technical Standards
(1) Ensure reliably grounded. (2) Ensure good contact on the clamping surface, the test current should not be less than 100 A. (3) Remove paint and metal oxide layers from the contact connections surfaces. (4) Reduce the length of the cable and use the cable with a sufficiently large cross-sectional area. (5) Prevent sudden tripping of the circuit breaker or sudden disconnection of the measurement circuit	The "JJG 1052–2009 Loop Resistance Calibration Regulations," national standards GB 763, GB 50150, and power industry standard DL/T 596 adopt the DC voltage drop method for measuring the conductivity loop resistance of circuit breaker, with a current no less than 100 A

(Ⅶ) Results Judgment	(Ⅷ) Digital Resources
(1) After major repairs and during handover acceptance, the loop resistance should comply with the manufacturer's specifications. (2) During operation, the loop resistance should not exceed 1.2 times the factory value. (3) Switches: The technical requirement for the loop resistance of a 1,250 A switch is ≤42 μΩ, and for a 1,600 A switch, it is ≤35 μΩ	Circuit breaker conductive circuit resistance test Disconnecting switch conductive circuit resistance test

Module 2 Follow Me

The difference between loop resistance and DC resistance lies in the electrical equipment and the range of test currents used. A loop resistance tester can measure within the range of 100–600 A, while a DC resistance tester typically measures at the position of 1/3/5/10/30 A. DC resistance is mainly used in transformer testing, while loop resistance is primarily applied to devices such as circuit breakers and disconnect switches.

Ⅰ. Purpose of Loop Resistance Testing

The conductive circuit of high-voltage switchgear is composed of multiple conductors that allow electrical conduction through contact. The DC resistance of the conductive circuit includes not only the DC resistance of the conductor material itself but also the contact resistance of various electrical contacts. The conductive circuit of Gas Insulated Switchgear (GIS) is even more complex. The resistance of the conductor itself depends on the resistivity of the conductor material and its geometric dimensions.

In the long-term operation of electrical switch equipment, the contact resistance is affected by chemical corrosion, mechanical wear, electrical wear, short circuit contact welding and other factors. If the contact resistance is too large, the heat generated under the long-term operating current will increase with the increase in resistance, the temperature of the electrical contact rises sharply, which may result in a decline in the insulating properties of the insulating material and mechanical strength, The oxidation of the contact surface leads to further deterioration. In severe cases, it may cause partial welding of the contact, affecting the normal opening and closing of the switch. When passing the short-circuit current, it will also affect the dynamic, thermal stability and interruption performance of the switch.

If the contact resistance is too high, the heat generated by the working current will increase with the increasing resistance over time. The temperature of the electrical contact will sharply rise, which may cause the insulation performance and mechanical strength of the insulating material to decrease, and the contact surface to oxidize further, resulting in more severe deterioration. In severe cases, local welding of the contact may occur, which can affect the normal switching of the switch. When short-circuit current flows through, it can also affect the dynamic and thermal stability performance and breaking performance of the switch.

Contact resistance is affected by four main factors: material properties (aluminum, copper, silver, tin), contact form (point contact, line contact, surface contact), contact pressure, and processing technology of contact surfaces. Therefore, current and resistance measurements of the conductive circuit are a standard testing item in type tests, factory tests, handover acceptance tests, and preventive tests of high-voltage switches. The resistance value of each phase's conductive circuit of electrical switchgear is an important data for installation, maintenance, and quality acceptance. By measuring the contact state of electrical contacts, safe operation of the switchgear can be ensured. Take Figure 3-6 as an example.

Figure 3-6 35 kV circuit breaker and disconnect switch

The loop resistance test is to measure the contact resistance of the conductive circuit of the switch when it is in the closed position. It checks for any contact defects, whether the circuit is well-connected, and if there is any oxidation on the contact surface. The resistance of the conductive circuit of the switchgear mainly depends on the contact resistance between the moving and static contacts of the circuit breaker (disconnect switch). The presence of contact resistance increases the loss of the conductor during power transmission, raising the temperature at the contact point. The magnitude of the resistance directly affects the current-carrying capacity during normal operation, and to some extent, the ability to interrupt short-circuit currents.

II. Conductive Circuit Test in High Voltage Switch

Electrical switchgear requires that the temperature rise does not exceed a certain value when the electrical contact passes through the rated current for a long time, and the contact resistance must be stable. When a short-circuit current passes through the switchgear for a short period, it is required that the electrical contact does not result in welding or splashing of contact materials. During the closing and opening process of the switch, the contact should be able to handle the short-circuit current without causing welding or severe damage. During the opening process, it is required that the contact wear should be minimized as much as possible when breaking the current.

The DC resistance value of the conductive circuit in a high-voltage switch is determined through type testing of the product. According to the GB/T 11022 standard, DC resistance measurement should be conducted before and after temperature-rise tests. The measurement should be taken under the same conditions as the test cycle's temperature of the test conductor and ambient air temperature. The difference in resistance between the two measurements should not exceed 20% of the resistance value obtained before the temperature-rise test.

This regulation requires that the DC resistance value must be determined by actual measurements to ensure long-term current-carrying capacity and performance when passing through extreme short-time currents. When measuring DC resistance during factory tests, it is necessary to obtain a measurement as similar as possible to the type test conditions (ambient air temperature, location of the measurement). The measured resistance should not exceed 120% of the resistance obtained before the temperature-rise test.

The measurement method mainly adopts the direct current voltage drop method. A certain direct current (generally not less than 100 A) is passed through the conductive circuit, and the voltage drop of the conductive circuit is measured using a direct current voltmeter. The direct current resistance of the conductive circuit is then calculated using Ohm's Law. Since the current passing through the test specimen is relatively large, it is sufficient to break down the metal oxide film on the contact surface, thereby reducing measurement errors. The data obtained from the millivoltmeter in the circuit resistance meter is relatively accurate. The principle of the direct current voltage drop method is shown in Figure 3-7.

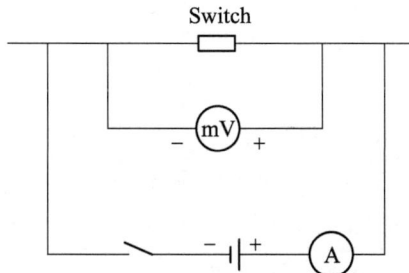

Figure 3-7　Schematic diagram of the direct current voltage drop method

When measuring the DC resistance of the conductive circuit in open-type high-voltage switchgear such as vacuum circuit breaker, SF$_6$ circuit breaker, and disconnect switch, please note the following: The voltage leads should be connected near the contact side. The current leads should be connected to the outside of the voltage leads. Ensure good contact between the voltage and current leads. If necessary, polish the contact surface with sandpaper. Before measuring, the switch should be operated several times at its rated operating voltage and gas pressure (or oil pressure) to ensure good contact with the contacts. This allows the measurement results to reflect the actual conditions. The short-circuit breaking currents that the SF$_6$ circuit breaker can handle are relatively large. The nameplate of a 220 kV SF$_6$ ZF16-252 GCB circuit breaker is shown in Figure 3-8.

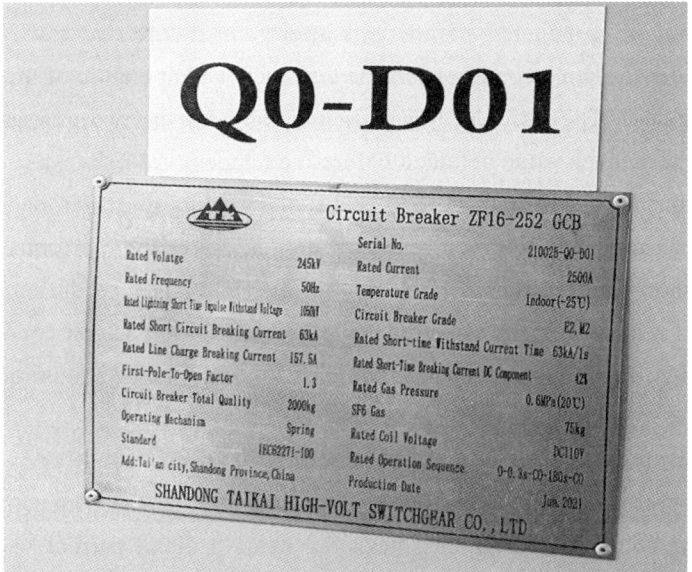

(a) 220 kV SF$_6$ Circuit Breaker ZF16-252GCB Nameplate

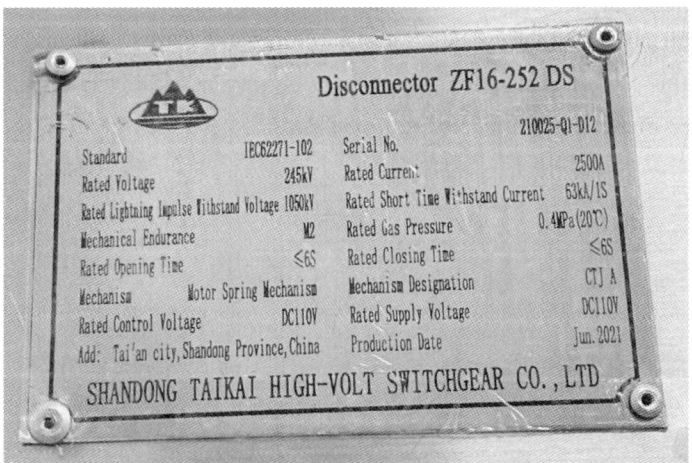

(b) 220 kV Disconnector Zf16-252DS Nameplate

(c) 220 kV Earth Switch ZF16-252 ES Nameplate

Figure 3-8   The Nameplate

The wiring diagram for the loop resistance test is shown in Figure 3-9.

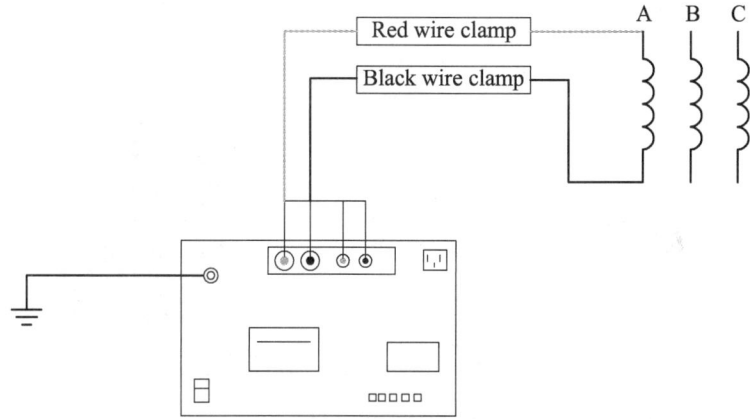

Figure 3-9   Wiring Diagram for Loop Resistance Test

The test procedure is shown in Table 3-3.

Table 3-3   Preventive Test Procedure for Conductive Circuit Resistance

Conductive Circuit Resistance	(1) During handover acceptance. (2) During annual preventive tests in spring. (3) After major maintenance. (4) When necessary	(1) After major maintenance and during handover acceptance, it should comply with the manufacturer's regulations. (2) During operation, it should not exceed 1.2 times the factory value. (3) For switches: The technical requirement for the loop resistance of a 1,250 A switch is ≤42 μΩ, and for a 1,600 A switch is ≤35 μΩ	If measuring using the direct current voltage drop method, the current should not be less than 100 A

Notes: ① When measuring the loop resistance, the contact springs should not be subjected to high currents.
② The current and voltage lead clamps should have good contact.

Module 3 Workshop

I. Method for measuring the main loop resistance of GIS

The main loop resistance of GIS is mainly measured using the grounding switch circuit. If there is insulation at the connection point between the grounding switch and the shell, the measurement can be taken by opening the connection. If there is no insulation at the connection point between the grounding switch and the shell, first measure the resistance R_1 of the GIS shell, then measure the resistance value R_2 after the circuit is in parallel with the shell, and then calculate the main loop resistance $R=R_1R_2/(R_1-R_2)$. For a sleeve with overhead lines entering it, the measurement current can also be injected from the sleeve side, and the measurement points should be as dense as possible.

Figure 3-10 shows a high-precision circuit resistance tester.

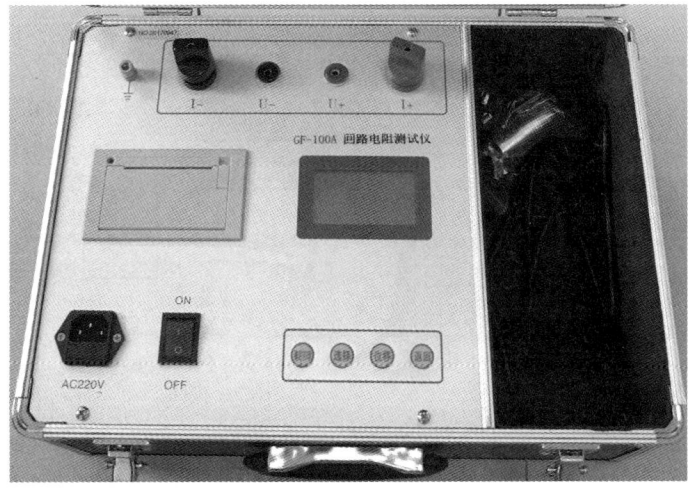

Figure 3-10 High Precision Loop Resistance Tester

The wiring principle for measuring the GIS loop resistance is shown in the figure 3-11.

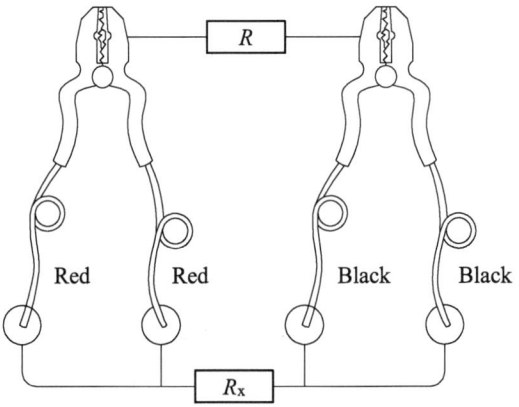

(a) Schematic diagram of resistance measurement wiring

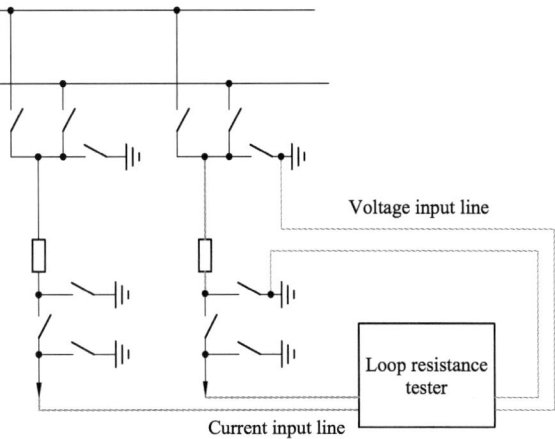

(b) Wiring diagram for injecting current from the sleeve and measuring voltage from the grounding switch

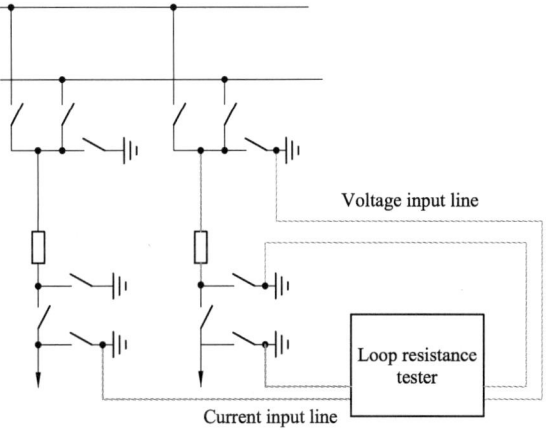

(c) Wiring diagram for injecting current from the grounding switch and measuring voltage from the grounding switch

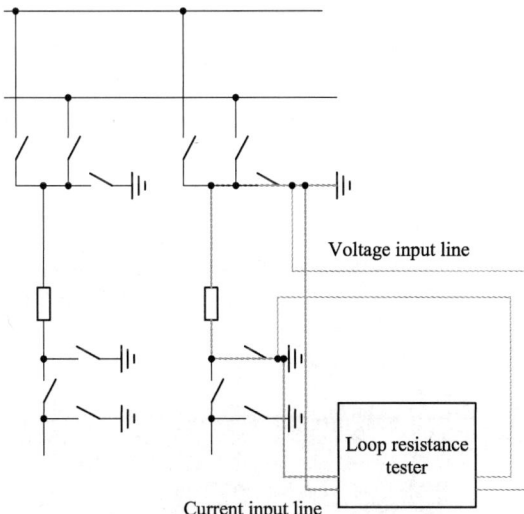

(d) Wiring diagram for injecting current and measuring voltage at the same location

Figure 3-11　The wiring principle for measuring the GIS loop resistance

Ⅱ. Loop Resistance Measurement

After correct wiring, the measurement main interface will be displayed when the device is turned on. This is shown in Figure 3-12.

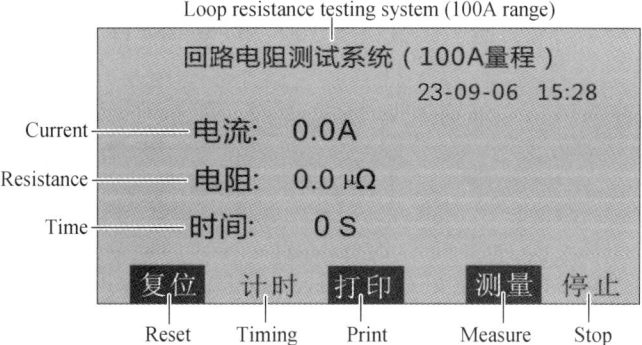

Figure 3-12　Measurement main interface

Click on the "Timer" button to set the measurement time. This is shown in Figure 3-13.

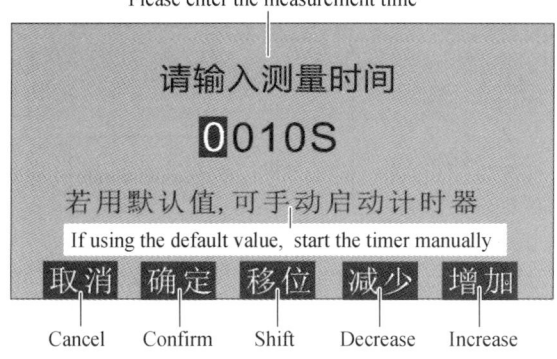

Figure 3-13　Measurement time setting menu

Click the "Measure" button to start the measurement. The measured result is 20 μΩ, indicating that the data is qualified. This is shown in Figure 3-14.

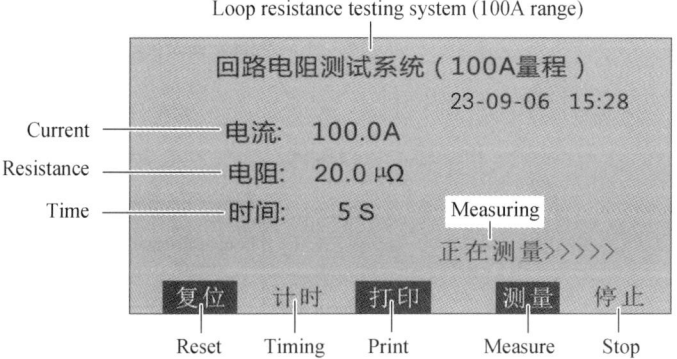

Figure 3-14　Measurement results

The loop resistance test procedure is shown in Table 3-4.

Table 3-4 Loop Resistance Test Procedure

Test Item	Routine Test Standard (Reference Cycle: 3 years)	Handover Acceptance Test Standards
Measurement of Main Loop Resistance	Difference with Initial Factory Value < 30%	The test results should comply with the specifications of the product's technical requirements

Test Notes:

(1) Apply any value of DC current between 100 A and rated current in the circuit and maintain it for 1 minute.

(2) Measurement should be performed using instruments that reflect average values, with a precision of not less than 0.5 grade for measuring meters.

(3) The voltmeter is connected on the inner side of the tested circuit and is switched on after the power circuit is connected, to prevent measurement interruption caused by sudden tripping of the circuit breaker or sudden disconnection of the measurement circuit that may result in damage.

(4) If the infrared thermal imaging displays abnormal temperatures at fracture surfaces, abnormal temperature differences between phases, or if there have been more than 100 switching operations of the circuit breaker after the test, this test should also be conducted.

(5) The loop resistance testing instrument should be reliably grounded. Ensure good contact on the clamping surface of the measurement cable, and the test current should not be less than 100 A. Remove paint and metal oxide layers from the contact surfaces of the tested equipment's terminal connections. Perform several opening and closing operations on the circuit breaker before the test to minimize the impact of the conductive circuit oxide film on the test results.

It is not allowed to energize when the test cable is open-circuited, as it may damage the instrument. To minimize voltage drop in the measurement cable, try to reduce the length of the cable and use the cable with a sufficiently large cross-sectional area.

(6) During the measurement, prevent sudden tripping of the circuit breaker or sudden disconnection of the measurement circuit (such as cable detachment).

(7) When measuring the main circuit resistance of a vacuum switch, do not clamp the current line on the switch contacts spring to prevent damage to the spring.

When the measured values exceed the technical requirements specified by the manufacturer or when there is a significant difference between the three-phase comparison, the following reasons should be checked:

(1) The surface of the contact points may have oxidation or damage.

(2) There could be residual mechanical particles or carbides between the contact points.

(3) The pressure of the contact points may decrease due to mechanical reasons, such as mechanical sticking of the mechanism, rupture of the contact spring, annealing, etc.

(4) The effective contact area of the contact points may be reduced due to improper adjustment.

(5) Screws at the connection points of the conductive circuit may become loose.

Module 4　Training

1. What defects can be found by testing the loop resistance of a circuit breaker?

2. What is the purpose of testing the loop resistance of a circuit breaker?

Worksheet 3 Transformer Ratio and Grouping Measurement

Module 1 Operating Worksheet: Transformer Ratio Grouping Test

(Ⅰ) Test Name and instrument	(Ⅱ) Test Objects
Ratio and grouping test **Multifunctional Ratio and Grouping Testing Device**	Measure the turns ratio and wiring group of single-phase, three-phase transformer, and instrument transformer
(Ⅲ) Test Purpose	(Ⅳ) Measurement Steps
(1) Check if the tap voltage ratio of the transformer winding is qualified. (2) Verify if the turns ratio matches the nameplate specifications. (3) Determine if the leads and connections of each tap on the winding are correct. (4) After a transformer failure, check for inter-turn short circuits in the winding. (5) Determine if the transformer is suitable for parallel operation	(1) Connect the test leads to the high and low-voltage windings as required. (2) Set up the testing parameters on the instrument after powering it on. (3) Press the test button, complete the testing process, and record the test data. (4) Save or print the data, then disconnect the power supply. (5) Ensure proper discharge of the tested equipment. (6) Remove the test leads
(Ⅴ) Precautions	(Ⅵ) Technical Standards
(1) Before connecting the test leads, ensure that the transformer is fully discharged. (2) Connect the test leads, test specimens, and instruments according to the test requirements, and make sure the connections are secure and reliable. (3) During the measurement process, it is strictly forbidden to touch the test specimen	For the transformer with a voltage of less than 35 kV and a voltage ratio of less than 3, the allowable deviation for the voltage ratio is ±1%. For all other transformers, the allowable deviation for the rated tap voltage ratio is ±0.5%. The voltage ratios of other taps should be within 1/10 of the impedance voltage value of the transformer in percentage, but not exceeding ±1%

Continued

(Ⅶ) Result Judgment	(Ⅷ) Digital Resources
The voltage ratio of each tap on the transformer should not have significant differences compared to the nameplate value. For voltage ratios greater than 3, the error should be less than 0.5%. For voltage ratios less than or equal to 3, the error should be less than 1%	**Transformer turns ratio and vector group test** **Fully automatic transformer turns ratio test**

Module 2　Follow Me

Ⅰ. Introduction of Transformers

The voltage ratio and connection group indicated on the nameplate of a transformer are important parameters and essential conditions for the parallel operation of the transformer. If the connected transformer in parallel operation has inconsistent connection groups, circulating currents may occur between the two transformers after operation. The connection group of windings should be measured and compared with the nameplate value during the production, manufacturing, factory testing, handover acceptance, and overhaul of transformers. The railway traction transformer and its nameplate are shown in Figures 3-15, 3-16.

Figure 3-15　Exterior of three-phase 220 kV railway traction transformer

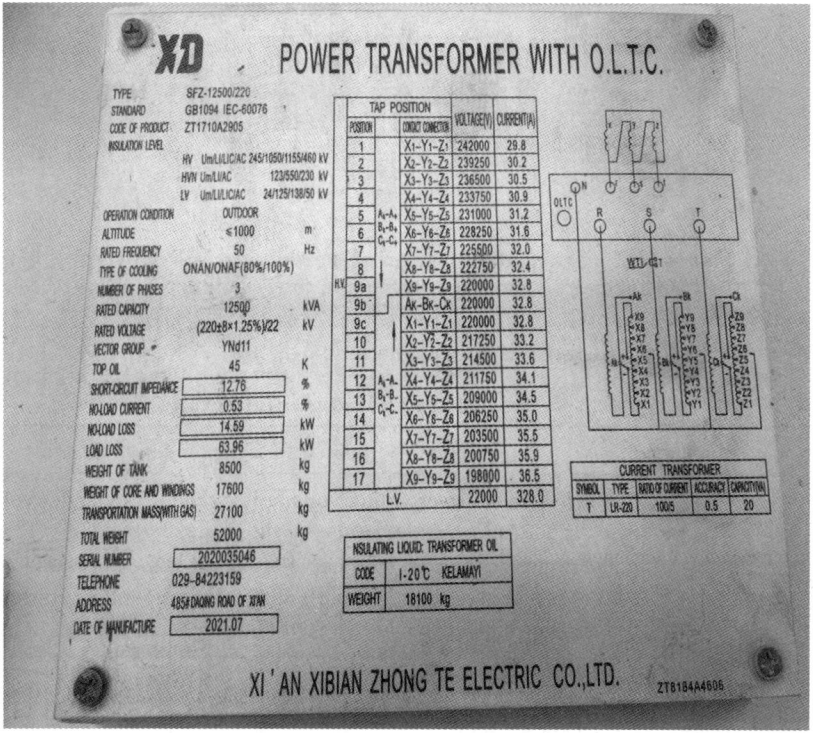

Figure 3-16   Nameplate of three-phase 220/20 kV power transformer

II. Transformer Voltage Ratio Test

The voltage ratio is an important performance indicator of a transformer. The purpose of measuring the voltage ratio of a transformer is as follows:

(1) Ensure that the voltage ratio of each tap of the winding is within the allowable technical range.

(2) Verify the correctness of the winding turns and check if the turn ratio matches the nameplate to ensure the correct voltage transformation.

(3) Determine whether the connections of the lead wires and tap switches for each tap of the winding are correct.

(4) Identify whether there are short circuits between layers or turns in the transformer winding.

(5) Provide basis for determining whether the transformer can be put into operation or operated in parallel.

Under the condition of no-load operation of the transformer, the ratio of the voltage at the high-voltage winding side (AX) to the voltage at the low-voltage winding side (ax) (as shown in Figure 3-17) is called the voltage ratio of the transformer.

$$K = \frac{U_1}{U_2}$$

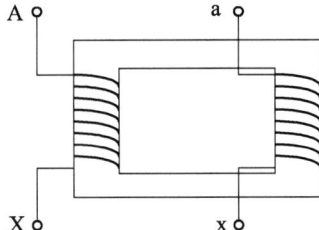

Figure 3-17 Schematic diagram of the voltage ratio between the transformer's primary winding and the secondary winding

Qualified standards for turns ratio is shown in Table 3-5.

Table 3-5 Qualified standards for turns ratio

ID	Power Equipment	Qualified standards
1	Transformer	For the transformer with a voltage less than 35 kV and a voltage ratio less than 3, the allowed deviation of the voltage ratio is ±1%. For all other transformers with rated tapping voltage ratios, the allowable deviation is ±0.5%, and the voltage ratio of other taps should be within 1/10 of the transformer impedance voltage, but not exceed ±1%
2	Instrument Transformer	The voltage ratio of the transformer joint should not differ significantly from the nameplate value. When the turns ratio is greater than 3, the error should be less than 0.5%, and when the turns ratio is less than or equal to 3, the error should be less than 1%

III. Transformer Groups

The transformer connection group refers to the phase difference between the line potential of the secondary winding and the corresponding line potential of the primary winding. It is related to the winding arrangement, the direction of the marked ends of the winding wires, and the connection method of the three-phase windings, as shown in Figure 3-18.

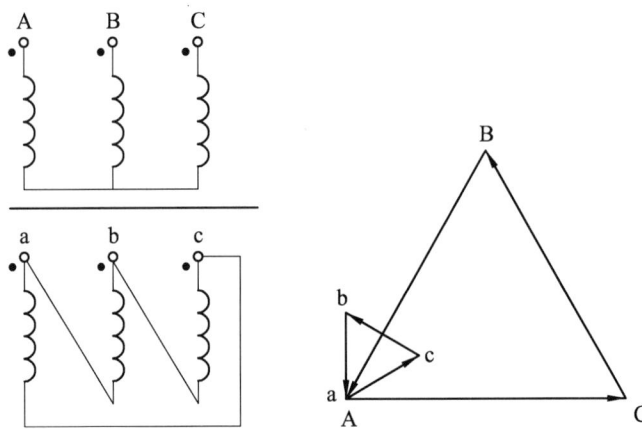

Figure 3-18 Transformer connection groups

The phase angle between the line voltages of the primary and secondary sides of a three-phase transformer depends on the coil connection method. There are twelve different angles at which the secondary line voltage lags behind the primary line voltage, namely 30°, 60°, 90°, 120°, 150°, 180°, 210°, 240°, 270°, 300°, 330°, and 360°. These correspond to twelve different connection points, labeled as 1, 2, 3, 4, 5, 6, 7, 8, 9, 10, 11 and 12, respectively.

Common standard groupings include: Yyn0 for the three-phase power transformer used in three-phase four-wire distribution systems supplying mixed loads for power and lighting; Yd11 for the three-phase power transformer used in low voltage lines above 0.4 kV; YNd11 for the three-phase power transformer used in high voltage lines above 110 kV where the neutral point needs to be grounded; YNy0 for the three-phase power transformer used in systems where the primary side needs to be grounded; Yy0 for the three-phase power transformer used in three-phase power load lines.

The commonly used main connection groups are shown in Table 3-6.

Table 3-6 Characteristics and application ranges of several commonly used coil connections

Connection Method	Characteristics and Application Range
Y/Y	(1) The coil winding has a large wire cross-section, which allows for high space utilization of the coil. It is used in the distribution transformer and can also be used in the interconnecting transformer or special transformer with a symmetrical three-phase load. (2) The neutral point can be brought out to supply three-phase four-wire loads. however, for the three-phase group composed of the single-phase transformer (hereinafter referred to as the single-phase group) or the three-phase three-column side-yoke core transformer (hereinafter referred to as the three-column side-yoke type), the neutral point of the primary side must be connected with the neutral point of the power supply, otherwise, this connection method cannot be adopted. (3) For the three-phase three-column core transformer (referred to as the three-column transformer), the neutral point of the primary side should not be connected to the neutral point of the power supply. When supplying three-phase four-wire loads on the secondary side, the line current should be limited
△/△	(1) The coil winding has a small wire cross-section, resulting in low space utilization of the coil. It is only suitable for the low voltage and high current transformer. (2) It allows for asymmetrical three-phase loads. When one phase fails, the remaining two phases can continue to operate by connecting them in a V configuration. At this time, the three-phase output capacity is reduced to $1/\sqrt{3}$ (For the three-phase transformer, the coil of the faulty phase must be disconnected and opened from the rest two phases. If the fault is due to inter-turn short-circuit, it cannot be reconnected in a V configuration and continue operation). (3) It does not produce third harmonic voltages, but it cannot supply three-phase four-wire loads and is also not suitable for high-voltage transformers
Y/△ or △/Y	(1) It does not produce third harmonic voltages and is suitable for various large and medium-sized transformers. When the △/Y connection method is used for distribution transformers, it allows for a greater degree of asymmetry in three-phase loads compared to the Y/Z connection method, and the neutral current can reach about 75% of the rated current. However, the lead structure is more complex, and the disadvantages of the △ connection method are the same as the first point of the △/△ connection method. (2) The neutral point of the Y connection method can be brought out. (3) If a coil of any phase fails, the transformer must be stopped

Continued

Connection Method	Characteristics and Application Range
Y/Z	(1) The neutral point can be brought out and can supply three-phase four-wire loads. It is suitable for distribution transformers or special transformers and allows for a greater degree of asymmetry in three-phase loads compared to the Y/Y connection method. The neutral current can reach about 40% of the rated current. (2) There is no third harmonic component in the phase voltage of the Z connection method. (3) Compared with the coil of the Z connection method, the coil of the Z-Y connection method requires 15.5% more wire, and it is only suitable for low-voltage windings
△/Y	(1) It is the same as the first and third points of the △/Y connection method, but only suitable for distribution transformers or special transformers. (2) It is the same as the third point of the Y/Z connection method

According to the "Standard for Electrical Equipment Handover Acceptance Test in Electrical Installation Engineering" (GB 50150–2016): When the tap changer leads are disassembled or the winding is replaced, the turns ratio and connection group of the transformer must be tested.

(1) Check the voltage ratio of all tap joints. Compared with the manufacturer's nameplate data, it should comply with the "no significant difference" rule and follow the rule. The "no significant difference" rule is as follows: For the transformer with a voltage of less than 35 kV and a voltage ratio of less than 3, the allowable deviation of the voltage ratio is ±1%; for all other transformers, the allowable deviation of the rated tap voltage ratio is ±0.5%, and the voltage ratio of other taps should be within 1/10 of the transformer's impedance voltage value, but not more than ±1%.

(2) Check the three-phase connection group of the transformer and the polarity of the outgoing wires of the single-phase transformer, which must match the design requirements and the markings on the nameplate and symbols on the casing.

(3) Test procedure for voltage ratio status check of transformer winding taps in power transmission and transformation equipment.

① The initial value difference should not exceed ±0.5% (rated tap position) or ±1.0% (other position) (warning value).

② This test should be conducted when the core components or the main body are disassembled for maintenance or when there is suspicion of winding defects. The test result should be consistent with the nameplate markings.

The nameplate of a dry-type power transformer is shown in Figure 3-19.

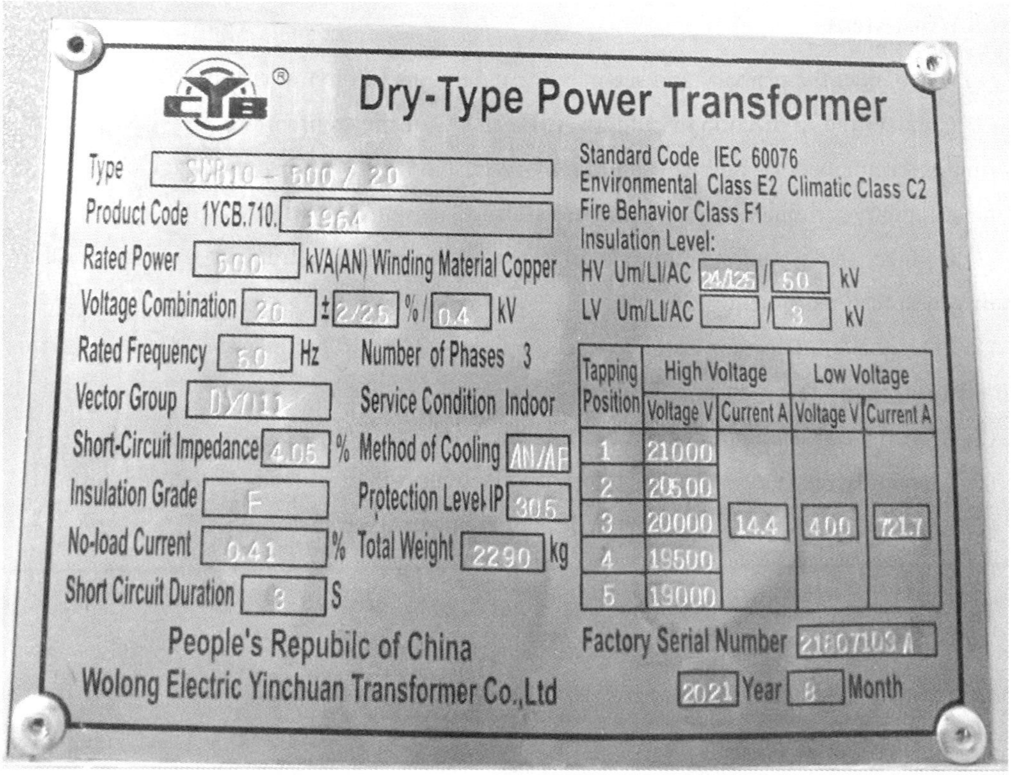

Figure 3-19 Nameplate of dry-type power transformer

Module 3 Workshop

Ⅰ. Transformer ratio group testing

(Ⅰ) Transformer ratio wiring diagram (see Figure 3-20)

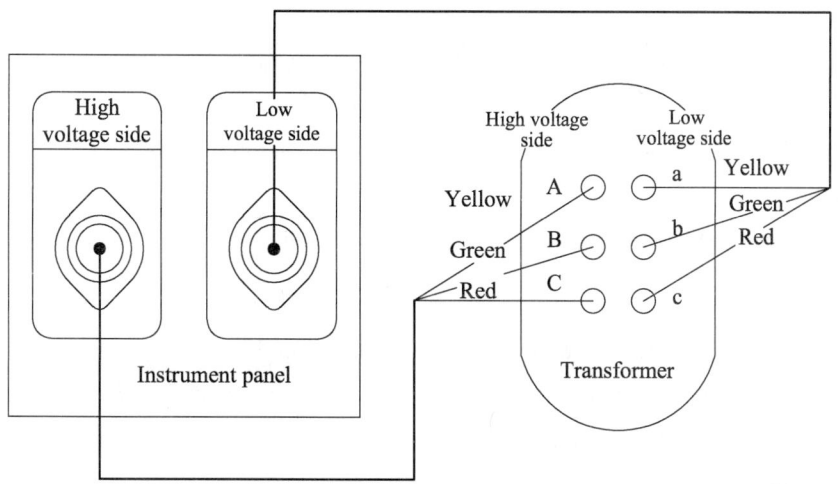

Figure 3-20 Wiring diagram of transformer ratio testing instrument

(Ⅱ) Test steps

(1) Disconnect the primary and secondary connections of the transformer.

(2) Connect the high-voltage outputs A, B, and C of the transformer ratio test instrument to the primary terminals X, Y, and Z of the transformer, and connect the low-voltage outputs a, b, and c to the secondary terminals x, y, z of the transformer, as shown in Figure 3-20.

(3) Connect the grounding terminal first, then connect the instrument terminals, and ground the turns ratio test instrument.

(4) Enter the "transformer group" "total tap numbers" "differential" and "rated turns ratio" on the transformer ratio test instrument.

(5) After the test is completed, record the test data.

(6) After fully discharging the transformer, disconnect the test leads of the transformer.

(7) Restore the primary and secondary connections of the transformer.

(8) Clean up the site.

The instrument is shown in Figure 3-21.

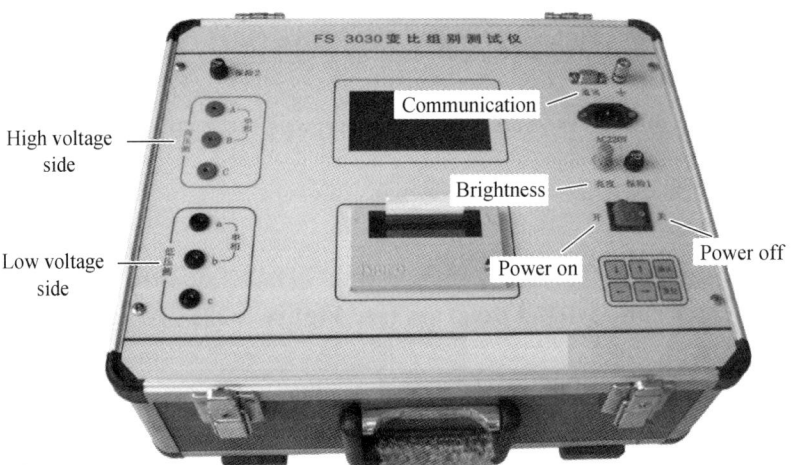

Figure 2-21　Fully automatic ratio grouping tester

Ⅱ. Test operation method

(1) Connection.

Turn off the power switch of the instrument. Connect the high-voltage terminals A, B, and C of the instrument to the high-voltage side of the transformer, and connect the low-voltage terminals a, b, and c to the low-voltage side. The single-phase and three-phase connection methods are shown in Table 3-7.

Table 3-7 Single-phase and three-phase connection methods

Single-phase Transformer		Three-phase Transformer	
Instrument Terminals	Transformer Terminals	Instrument Terminals	Transformer Terminals
A	A	A	A
B	X	B	B
C	No connection	C	C
a	a	a	a
b	x	b	b
c	No connection	c	c

The neutral point of the transformer is not connected to the instrument or to the ground. Make sure to connect the instrument's grounding wire.

(2) Turn on the power switch of the instrument and enter the wiring method settings. Based on the actual nameplate information of the transformer, select Yy, Y/d, Dd, and D/y for the single-phase transformer or three-phase transformer, and select Dy for the three-phase transformer when the actual wiring method is unknown. Set the turns ratio according to the actual nameplate information of the transformer. For example, for a 10 kV/0.4 kV transformer, enter a turns ratio of 25 and then press confirm to save the data and exit. Press the measurement button to start the measurement. After the measurement is completed, the measurement result will be displayed as shown in Figure 3-22.

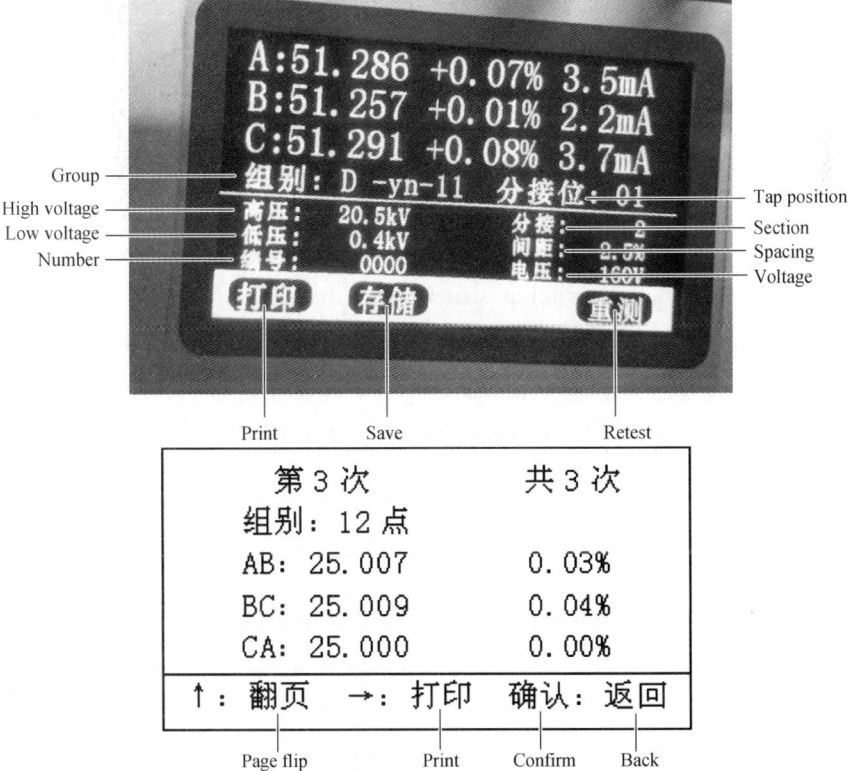

Figure 3-22 Test results for the transformer ratio group

As shown in the figure 3-22, the left side represents the turns ratio for phases AB (values of 25.007, 25.009, 25.000), and the right side represents the relative error for phases AB (values of 0.03%, 0.03%, 0.00%). If the measured relative error of the turns ratio is greater than 10%, it will be displayed as ">10%". If the measured relative error of the turns ratio is less than −10%, it will be displayed as "< − 10%".

III. Test Result Analysis

Based on the test results, compare them with relevant standards, factory data, and test data of similar equipment. According to the "DL/T 596–2021 Pre-commissioning Test Code for Power Equipment" regulations.

IV. Precautions

In order to use the testing instrument correctly and obtain accurate turns ratio error, the following issues should be taken into account during the testing process. When connecting the transformer to the testing instrument, make sure that the high and low-voltage sides are not reversed. If the "high/low voltage side reversed" indicator on the testing instrument lights up during the test, promptly turn off the instrument, check and correct the wiring, and then conduct the test. Otherwise, high voltage may enter the bridge. During the test, it is not allowed to touch the terminal connections of the transformer and the testing instrument. Before the test, input the transformer group based on the nameplate values, and the instrument will automatically calculate the relative error ratio between the actual measurement value and the nameplate value.

The transformer ratio should be measured at each tap position. For all tap switch positions, separate measurements need to be taken. Pay attention to whether the tap switch connections correspond correctly to the nameplate values. Check if the instrument settings match the transformer's actual tap position. If there is a voltage regulating device, use an electrically operated device to change the tap switch position before measurement.

When measuring the groups, for the transformer with a high turns ratio, choose a higher voltage and a smaller range DC millivoltmeter, microammeter, or multimeter. For the transformer with a small turns ratio, choose a lower voltage and a larger range millivoltmeter, microammeter, or multimeter.

Module 4　Training

What are the qualification criteria for the transformer ratio?

_____
_____
_____

Worksheet 4　Transformer Winding Deformation Test

Module 1　Operating Worksheet: Winding Deformation Test

(Ⅰ) Test Name and Instrument	(Ⅱ) Test Object
Transformer winding deformation test 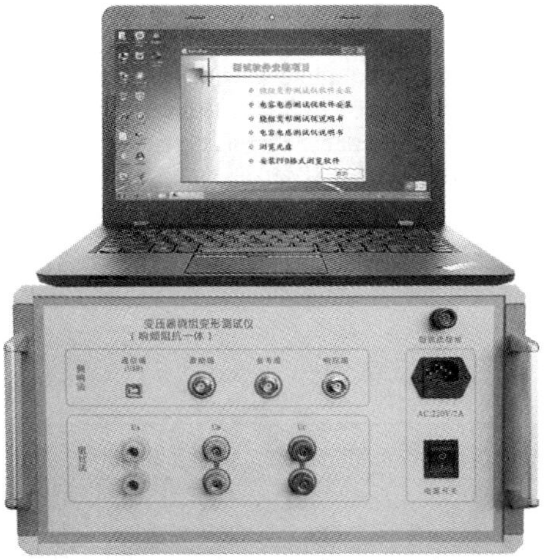 **Transformer winding deformation Instrument**	Various types of transformer winding testing
(Ⅲ) Test Purpose	(Ⅳ) Measurement Steps
The purpose of the transformer winding deformation test is to examine whether the winding of the transformer has experienced distortion, bulging, displacement, inter-turn short circuit, and displacement of the core body due to mechanical or electrical forces during transportation and installation	(1) Disconnect the on-load tap changer and the cooling power supply of the transformer, remove the protection cover of the transformer body, and fully discharge the grounding of each winding of the transformer. Remove or disconnect all external connections. (2) Establish a storage path for the test data in the laptop and input various measurement information. ① The storage path for the measurement data should clearly reflect the installation position, operating number, test date, and other information of the tested transformer for easy retrieval and to prevent data loss. ② Establish a test database and input basic information such as the nature of the test, transformer tap position, nameplate information, ambient temperature and humidity, test date, and test personnel. (3) For different windings of the transformer, connect and measure each phase winding of the transformer according to the requirements of the testing instrument. (4) After completing the test, save all the measured data for subsequent analysis during operation

Continued

(V) Precautions	(VI) Technical Standards
(1) In the low-frequency part (tens of kHz), the frequency response curve generally overlaps well. Otherwise, a poor test connection should be suspected first. (2) Generally, the consistency of frequency response characteristics of the transformer of 35 kV and below may be poor, and the original data should be kept for comparison at handover acceptance. (3) The measured frequency response curve is generally between +20 dB to -80 dB. If it exceeds this range, check whether the test circuit has a poor contact or broken connection. (4) When conducting a separate test on the angular-connected winding, the three-phase frequency response characteristics may be inconsistent. Balanced windings may cause inconsistency in the three-phase frequency response characteristics. Severe deformation of the winding will affect the frequency response characteristics of adjacent windings. Temperature also has an impact on the frequency response characteristics	(1) "Power Safety Regulations of State Grid Corporation-Substation Part" (Q/GDW 1799.1 – 2013) (2) "Maintenance and Test Procedures for Transmission and Transformation Equipment" (Q/GDW 1168 – 2013) (3) "Test Specifications for Electrical Equipment Handover Acceptance in Electrical Installation Projects" (GB 50150 – 2016)
(VII) Result Judgment	(VIII) Digital Resources
The frequency response curve of the transformer winding in the low-frequency range (10–500 kHz) exhibits abundant resonance points. The variations of these resonance points sensitively reflect the deformations such as broken strands, bulges, distortions, and inter-disk displacement of the transformer winding. The high-frequency range (above 500 kHz) can indicate the displacement of the transformer winding. However, for the high-frequency part of the frequency response curve of the transformer winding at 110 kV and above, due to numerous influencing factors, it is sometimes difficult to ensure a good overlap of this part of the curve. When making judgments, particular pay attention to the low-frequency and middle-frequency parts, while the high-frequency part can serve as a reference when necessary	**Transformer Winding Deformation Test**

Module 2 Follow Me

I. Types of Transformer Faults

The transformer is one of the most important equipment in power systems. During transportation, the transformer may encounter unexpected collisions and impacts. When operating under fault conditions, the impact current on the transformer will subject its windings and mechanical structure to mechanical stresses, resulting in a certain degree of deformation in the windings and causing accidents during operation, as shown in Figure 3-23.

Figure 3-23 Transformer explosion and fire accident

Due to the severe harm that winding deformation poses to the operation of the transformer and power system, previous testing methods were unable to effectively detect such defects. Verification could only be done through hanging cover inspection, which not only required a significant amount of manpower and resources but also posed certain risks to the transformer itself. Moreover, in the current operation of power systems, it is challenging to have long-term power outages for the large transformer. Therefore, the frequency response method and low-voltage short circuit impedance method, which can quickly measure internal deformations of windings without the need for hanging cover inspection, have been widely used in on-site operations. The transformer structure is shown in Figure 3-24.

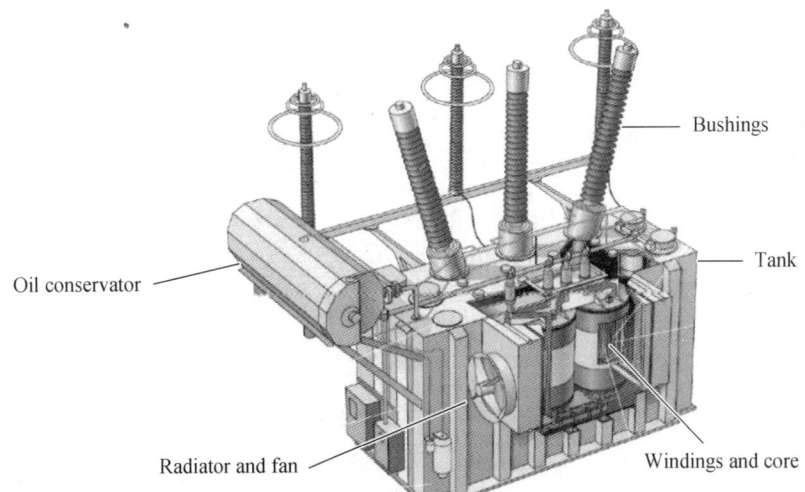

Figure 3-24 Transformer structure

The main reason for the deformation of the transformer winding are short circuit, near-area short circuit or multiple overcurrent operations, and collision during transportation. The main causes of transformer faults are equipment aging, improper maintenance, overload operation, and electrical faults. The three main types of transformer faults are concentrated on winding, tap switch, and bushing failures, which account for about 70% of all transformer faults. The reasons and types of transformer faults are shown in Figure 3-25. To analyze transformer faults, factors such as overheating (due to deteriorating transformer oil), bushing main body issues (such as partial discharge and moisture problems), electrical performance (such as corona discharge), and on-load tap switch (such as arcing) can be analyzed. These problems run through the entire lifespan of a transformer, always being subjected to the influences of multiple factors such as heat, electricity, mechanics, and chemistry.

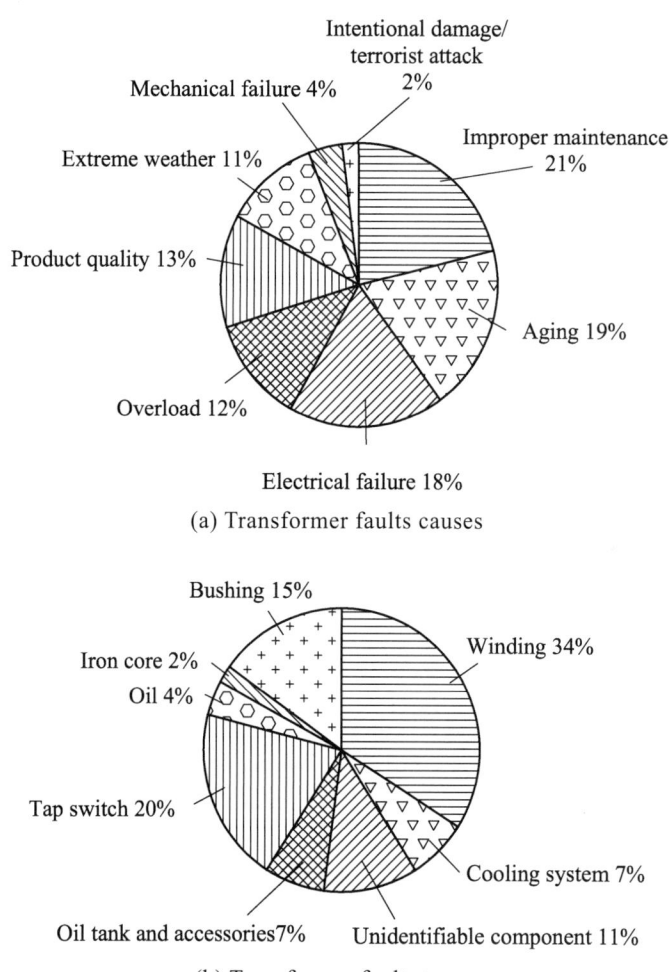

(a) Transformer faults causes

(b) Transformer faults types

Figure 3-25　The reasons and types of transformer faults

II. The external force on the transformer and deformation cause

The main components of the transformer are winding and core. To solve the problems of heat dissipation, insulation, sealing and safety, oil tank, insulating bushing, oil storage cabinet, cooling device, pressure release valve, safety airway, thermometer and gas relay are also needed. The normal appearance of the transformer and the winding is shown in Figure 3-26.

Figure 3-26   Transformer appearance and internal structure

Transformer winding deformation refers to irreversible changes in the axial or radial dimensions, including displacement of the body, distortion, bulging, displacement of windings, and short-circuit between turns, caused by mechanical forces or the impact of fault short-circuit currents on the internal coil winding during transportation or operation of the transformer. After the deformation of transformer windings, the insulation distance changes, leading to a decrease in both the mechanical and insulation performance of the windings, which is difficult to detect through insulation tests and transformer oil tests, resulting in latent faults.

Therefore, winding deformation poses a major risk to the safe operation of the power transformer. When exposed to lightning overvoltage or short-circuit accidents, the windings may not be able to withstand the tremendous electrodynamic forces and can be damaged, leading to breakdown of insulation between turns or layers of the transformer. This can cause sudden insulation accidents, and even insulation breakdown accidents may occur under normal operating voltage due to prolonged partial discharge effects.

1. Types of Impact Forces on Windings

(1) Electrodynamic forces during normal operation

During normal operation, electrodynamic forces are usually small. However, if there are defects in the winding during the manufacturing process, such as loose windings, uneven wires or burrs, entry of blocks into the bending area of commutation, insulation damage at commutation points, and uneven wedges, the vibration caused by electrodynamic forces can further amplify these defects. As a result, the winding may deform during normal operation. There is also a possibility of prolonged friction between wires and blocks, leading to insulation damage and discharge. Additionally, if the thermal stability of the winding is insufficient, winding deformation failures may occur during normal operation.

(2) Sudden Short-Circuit Electrodynamic Forces

During a sudden short-circuit, the short-circuit current can be several times to tens of times higher than the normal rated current. The electrodynamic forces experienced by the windings are directly proportional to the square of the current. Therefore, in the case of a short-circuit, the electrodynamic forces can be tens to hundreds of times higher than during normal operation. Although the duration of a short-circuit is short, the powerful surge of current subjects the transformer windings to significant and uneven electrodynamic forces. This is particularly true when a short-circuit occurs at the transformer's output or nearby, where the combination of high short-circuit current and relatively low short-circuit impedance results in even greater electrodynamic forces. These forces can cause various types of deformation in the windings, making them the primary cause of transformer winding deformation.

(3) Direct mechanical impact force

The transformer is often subjected to external accidental mechanical impact forces during manufacturing, transportation, installation, and maintenance processes. According to Newton's mechanics theorem ($F = ma$), the transformer casing will experience acceleration or deceleration in the same direction as the external force, altering its previous motion state. This can result in changes such as transitioning from rest to motion, increasing or decreasing velocity, or transitioning from motion to rest. The internal windings, due to inertia, will continue to maintain their original state. As a result, relative movement occurs between the casing and the windings, becoming the cause of winding deformation. Depending on the specific connection between the transformer windings and the casing, different types of deformation or displacement may occur in the windings.

When a transformer is subjected to mechanical or electromagnetic forces, whether the windings deform and to what extent they deform are primarily influenced by two factors:

① The ability of the transformer windings to withstand the impact force, which mainly depends on the material, structure, manufacturing process, and stress uniformity of the windings.

② The characteristics of the impact force applied to the windings, including its magnitude, duration, frequency, mode, and range. The leads, taps, inter-winding connections, intermediate connections, tap connections, internal welded joints, and gaps due to insufficient winding or compression in the transformer windings are all structurally weak points that are prone to deformation. Even the transformer of the same factory and specifications may vary due to manufacturing deviations, safety factor deviations, and other random probabilities.

In summary, if the mechanical strength of a transformer is insufficient to withstand powerful mechanical and electromagnetic forces, the windings will experience various types of deformation and displacement faults.

2. Causes of Transformer Winding Deformation

Common types of transformer failures include: sudden internal short-circuit faults inside the transformer caused by external short-circuit impacts, inter-layer and inter-turn short-circuit faults caused by the breakdown of the main insulation of a transformer due to overvoltage, and insulation damage caused by poor sealing of high-voltage bushings resulting in moisture ingress, as shown in Figure 3-27.

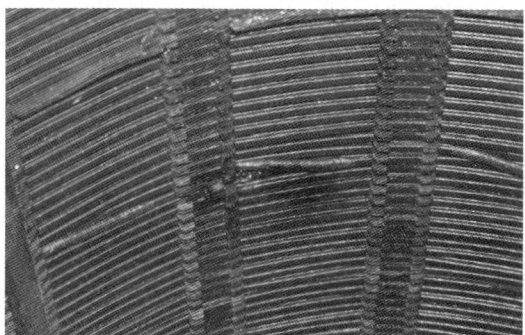

Figure 3-27　Transformer with inter-layer and inter-turn short circuit faults caused by lightning strike

The winding deformation test is a necessary item for the transformer during factory delivery, handover acceptance, and after experiencing short-circuit accidents. When a transformer suffers various short-circuit fault currents during operation, a large short-circuit current will flow through the transformer winding. Under the interaction with the leakage magnetic field, a large electrodynamic force will be generated. At this time, each winding will bear huge, uneven radial and axial electrodynamic forces, causing the coil to become deformed in a short time under the effect of heating and electrodynamic force, and even leading to interphase short-circuits and winding burnout.

Additionally, during transportation and installation, the transformer may suffer collision impacts that will cause mechanical displacement and deformation of the windings, leading to serious transformer accidents such as insulation damage, winding short-circuit, and burnout. Furthermore, the existence of dead zones or failure to act in the protective system will also cause the transformer to withstand the action of short-circuit currents for a long time, which is one of the reasons for the deformation of the winding.

External short circuits causing the deformation of transformer windings is a common fault during the transformer operation, seriously threatening the safe operation of the system. The transformer adopts semi-hard copper, self-adhesive transposition wire, winding coil with hard insulation cylinder, and inner and outer support bars of encryption coil to improve the anti-short-circuit ability of the transformer, which is based on the consideration of improving the anti-radial short-circuit ability.

III. Purpose and Methods of Transformer Winding Deformation Detection

1. Purpose of Transformer Winding Deformation Test

The transformer winding is the main part of transformer accidents, and the poor short-circuit resistance of transformer windings is the main cause of transformer operational damage. After the transformer winding deforms, it can still operate normally for a period of time. Since conventional electrical tests such as resistance measurement, ratio measurement, and capacitance measurement are difficult to detect winding deformation, this poses a serious threat to the safe operation of the power grid.

On the one hand, the deformed transformer is caused by changes in insulation distance or damage to the insulating paper. When encountering overvoltage, the winding may experience inter-layer or inter-turn breakdown. Alternatively, under the long-term working voltage, insulation damage gradually expands, ultimately leading to transformer damage. On the other hand, after the winding is deformed, its mechanical performance decreases. When it experiences a subsequent short circuit accident, it cannot withstand the tremendous impact force and may suffer damage. Therefore, testing and diagnosing transformers that have undergone mechanical and electromagnetic forces resulting in winding deformation is essential.

2. Methods for Detecting Transformer Winding Deformation

For newly installed transformers and repaired transformers, it is generally necessary to conduct winding deformation detection. Currently, it is common to use methods such as pre-factory inspection, post-installation inspection, routine inspections during operation, and comprehensive inspections after faults. By measuring and analyzing relevant characteristic parameters, it is possible to determine whether there are deformations, displacements, or other abnormal phenomena in the windings.

After the deformation of transformer windings, various abnormal phenomena are usually observed. Many characteristic parameters, such as electrical parameters, physical dimensions, geometric shapes, and temperature, show significant differences compared to normal conditions. Based on this, multiple methods for detecting winding deformation have been developed.

After experiencing a short-circuit impact, the transformer is generally inspected for its insulation condition using routine electrical test items and insulation oil analysis. The inspection results indicate that some transformers have passed the electrical tests and insulation oil analysis within the specified range of preventive test procedures. However, during the inspection of hanging cover, significant deformation of the windings or severe loosening of insulation pads was discovered. This indicates that routine electrical and oil tests are not effective in detecting winding deformation defects in transformers.

Although the inspection of the hanging cover is intuitive, it requires a significant amount of manpower and resources, and it is still difficult to determine if there is deformation in the inner windings. To compensate for the shortcomings of routine electrical methods and winding enclosure inspections, several mature detection methods are commonly used for detecting transformer winding deformation.

(1) Short-circuit Impedance Method

The principle of the short-circuit impedance method is to measure the current and voltage values in the transformer winding, calculate the short-circuit impedance value of the winding, and compare the change in the short-circuit impedance value before and after the deformation of the transformer winding. Based on the change in the leakage reactance value in the test transformer winding, the deformation or displacement of the winding can be determined. The short-circuit impedance of a transformer refers to the equivalent impedance at the input end of the transformer when the load impedance of the transformer is zero. It reflects the induced magnetic potential formed by the leakage magnetic flux between the windings or between the windings and the oil tank.

The short-circuit reactance component of the transformer is the leakage reactance of the transformer winding. At a certain frequency, the leakage reactance value of the transformer is determined by the geometric size of the winding. Any change in the structural state of the transformer winding will inevitably cause a change in the leakage reactance of the transformer. During measurement, the high-voltage side of the winding is connected to the power frequency AC source, and the low-voltage side is short-circuited. The wiring principle is shown in Figure 3-28.

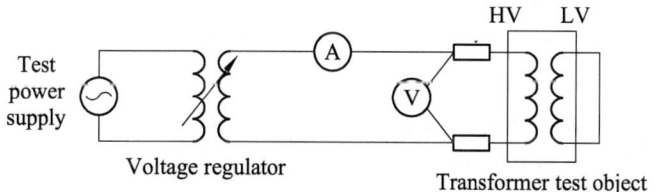

Figure 3-28 Wiring diagram for deformation testing of transformer winding using the short-circuit impedance method

(2) Low Voltage Pulse Method

Under the influence of higher frequency voltage, the magnetic permeability of the transformer core is nearly the same as that of air. The winding itself can be regarded as a passive linear distributed parameter network composed of linear resistors, inductors, capacitors, etc. Its equivalent circuit is shown in Figure 3-29, where L represents inter-layer inductance, K represents longitudinal capacitance, and C represents ground capacitance.

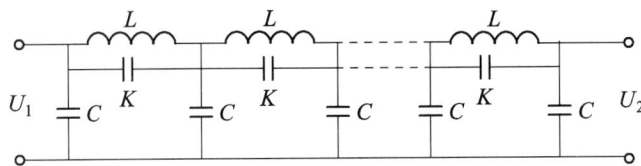

Figure 3-29 Equivalent circuit of transformer winding

The principle of the low voltage pulse method is to apply a stable low voltage pulse signal to one end of the transformer winding while recording the voltage waveforms at the terminal and other terminals simultaneously. By comparing the excitation and response in the time domain, a more accurate judgment can be made about the condition of the winding. When the winding of the transformer undergoes deformation, parameters such as inductance and capacitance in corresponding sections will change. Applying a pulse voltage excitation at the input end will cause changes in the response at the output end.

The low voltage pulse method is applied in on-site tests, but it is susceptible to various electromagnetic interferences during the testing process, resulting in poor repeatability. Moreover, it is not sensitive to faults in the first-end position of the winding, making it difficult to determine the location of winding deformation.

(3) Capacitance Variation Method

The principle of the capacitance variation method is that each winding of a transformer can be equivalent to a network composed of resistance, capacitance, inductance, and other elements. The equivalent capacitance of the winding directly reflects the relative position of the windings, the windings to the core, the windings to the tank and ground, and the structural characteristics of the windings. When the transformer is manufactured, the capacitance of each winding is relatively constant. As long as the transformer has not experienced short-circuit impacts, even under the influence of temperature and humidity, the capacitance variation is minimal.

After the transformer experiences a short-circuit impact, if the winding is not deformed or only slightly deformed, the change in capacitance is also small. However, if a certain winding is severely deformed, the capacitance variation will be significant. According to the "Preventive Testing Regulations for Power Equipment," the $\tan\delta$ of transformer windings should be tested every one to three years. During transformer handover acceptance and preventive testing, the $\tan\delta$ of each winding is measured, and the corresponding capacitance can be calculated. Therefore, based on the changes in capacitance of transformer windings, it is possible to determine whether the windings have undergone deformation.

This testing method is simple and convenient. However, due to the inherent scatter of the winding capacitances, it has poor sensitivity to faults such as bulges and distortions. It can be used as a supplementary test method for detecting winding deformations.

(4) Ultrasonic Reflection Method

The principle of the ultrasonic reflection method is to utilize an ultrasonic probe placed on the surface of the tested object to emit ultrasonic waves at a certain frequency. The ultrasonic waves propagate inside the tested object in longitudinal wave mode. When encountering an interface between two different media, reflection occurs, and the waves return along a certain path to be received by the ultrasonic probe.

By measuring the time, it takes for the ultrasonic waves to be emitted and received, the propagation time interval t of the ultrasonic waves within the tested medium can be obtained. For transformer windings and the steel walls of the casing, the distance between each point on the winding surface and the surface of the oil tank is a constant value. If there are abnormal faults such as indentation, protrusion, or displacement of the windings, the distance will change accordingly. By comparing the measurements, the deformation status of the windings can be determined.

When detecting the deformation of transformer windings using ultrasonic waves, the ultrasonic probe is placed on a certain position of the steel wall of the transformer casing, and a coupling agent (butter) is used to make the probe tightly contact the outer casing of the transformer and align the center of the probe with the winding to be measured. Under the action of a synchronous signal, the emission circuit excites the ultrasonic probe to emit ultrasonic waves. The ultrasonic waves pass through the steel wall and transformer oil and arrive at the transformer winding. Reflected waves are generated on its surface and returned along a certain path. Similarly, they pass through the transformer oil and the steel wall of the transformer casing and reach the ultrasonic receiving probe, generating a receiving electrical pulse signal. By processing with a related circuit, the time t taken for the ultrasonic waves to propagate and complete one round trip in the transformer steel plate and oil can be obtained.

The advantage of this method is that it is simple to operate, has good directness and repeatability. However, the disadvantage is that the results vary greatly in the presence or absence of oil, and the test results are easily affected by temperature.

(5) Vibration Method

The principle of the vibration method is to use a vibration sensor attached to the transformer tank to monitor the condition of the windings and iron core online. The characteristic vibration vectors of a well-performing transformer (including the spectra, power spectra, energy spectra, etc., of the winding and iron core vibration signals) are used as reference indicators. Once a fault occurs in the transformer winding, the changes in the current vibration characteristic vectors will be quickly reflected.

The advantage of the vibration method is that the testing system is not electrically connected to the entire power system, achieving online monitoring safely and reliably. Its drawback is that the power transformer can experience short-circuit faults at any time during operation. If a fault occurs in the internal winding of a suddenly short-circuited transformer, it will result in contact between the energized winding and the oil tank, which may have high voltage. In addition, transient induction can also generate high potential on the transformer body, affecting the safety of testing equipment and personnel.

(6) Frequency Response Analysis (FRA) Method

The working principle of the Frequency Response Analysis method is to detect the amplitude-frequency response characteristics of various windings of a transformer at different frequencies. After quantifying the measurement results, a transfer function characteristic curve of the transformer winding is generated. This allows for longitudinal comparisons of the same transformer at different periods or transverse comparisons of the transformer of the same type. By assessing the degree of change in the amplitude-frequency response characteristics, it is possible to determine if there is deformation in the transformer windings. In simple terms, a frequency is applied to the end of the transformer winding, and the response is measured at the other end. If there is a significant difference in the response, it indicates the presence of deformation; if the difference is small, it indicates good consistency.

At higher frequencies, the windings of a transformer can be equivalent to a passive linear two-port network composed of distributed parameters such as capacitance and inductance. The internal characteristics of this network can be expressed as a transfer function $H(j\omega)$. When the structure of the transformer is fixed, the relationship between the parameters and function curves of the transformer windings is determined. When changes occur internally in the transformer, the distributed parameters of the windings also change, resulting in corresponding changes in the function curves.

If the windings deform, the distributed inductance, capacitance, and other parameters inside the windings will inevitably change, causing the zeros and poles of the transfer function $H(j\omega)$ of the equivalent network to change. As a result, the frequency response characteristics of the network change. By establishing the functional relationship between the input excitation and output response of the instrument and plotting it point by point, the transfer function characteristic curve that reflects the characteristics of the transformer windings can be obtained.

The transfer function is the ratio of the output to the input of a passive two-port network expressed in Laplace transform form. The frequency response characteristics refer to the relationship between the transfer function $H(j\omega)$ of the network and the angular frequency ω under sinusoidal steady-state conditions. The variation of the amplitude of $H(j\omega)$ with ω is usually referred to as the magnitude-frequency response characteristic, and the variation of the phase of $H(j\omega)$ with ω is referred to as the phase-frequency response characteristic.

Sweep frequency detection involves continuously changing the frequency of the externally applied sinusoidal excitation signal source, measuring the ratio of the output signal to the input signal of the network at different frequencies, and plotting the corresponding magnitude-frequency response or phase-frequency response characteristic curves. In frequency response analysis, a stable sinusoidal sweep signal is applied to one end of the tested transformer winding, and the voltage amplitude and phase on that terminal and other terminals are simultaneously recorded, thereby obtaining a set of frequency response characteristics for the tested winding.

The magnitude-frequency response characteristic curve of the transformer winding typically contains multiple distinct peaks and valleys. The distribution position, number of peaks and valleys, and variations in amplitude can serve as important indicators for analyzing the deformation of the transformer windings. By using frequency sweep measurement on the characteristics of the transformer windings without removing the transformer cover or disassembling it, the amplitude-frequency response characteristics of each winding can be detected. For the transformer with voltages of 6 kV and above, the distortion, bulging, or displacement of the windings can be accurately measured, and the deformation of the windings can be automatically diagnosed by calculating relevant parameters from the curves.

The frequency response analysis method is mainly used to determine the deformation of transformer windings by longitudinally or transversely comparing the amplitude-frequency response characteristics of the windings, and comprehensively considering factors such as the transformer's short-circuit conditions, structure, electrical testing, and dissolved gas analysis in oil. Compared with the low-voltage pulse method, the frequency response analysis method avoids drawbacks such as bulky instruments and poor test repeatability, reduces the impact of electromagnetic interference, has good repeatability, and can analyze the frequency response curves more intuitively. It has high testing sensitivity and is widely used.

In conclusion, the short-circuit impedance method requires the use of large testing equipment, is time-consuming and labor-intensive, has low sensitivity, and is difficult to guarantee measurement accuracy, making it challenging for on-site applications. The low-voltage pulse method has poor repeatability over long intervals and is not sensitive to faults at the starting end of transformer windings. The capacitance variation method is greatly affected by the capacitance of the windings themselves, resulting in poor sensitivity for testing bulges, distortions, and other faults. The ultrasonic reflection method is significantly affected by oil temperature and the presence or absence of oil. The vibration method has an impact on the testing instrument as well as personal safety. The frequency response analysis method has good test repeatability, simple and lightweight testing equipment, high testing sensitivity, intuitive analysis of test spectra, and comparable data analysis, and is suitable for detecting deformation of transformer in operation in the power system.

Module 3　Workshop

Taking the frequency response analysis method as an example, the transformer deformation test is carried out as follows.

The transformer winding deformation tester is used to test the deformation of the power transformer with voltage levels of 6 kV and above and other special-purpose transformers.

Ⅰ. Transformer Winding Deformation Test Method

Apply a series of specific frequency signals to one end of each winding of the transformer and measure the response signals at both ends to obtain the frequency response characteristics. For windings with a neutral point lead, measure the frequency response characteristics of OA, OB, and OC in sequence. For delta-connected windings, measure the frequency response characteristics of AB, AC, and BC in sequence.

The wiring diagram for transformer winding deformation testing is shown in Figure 3-30.

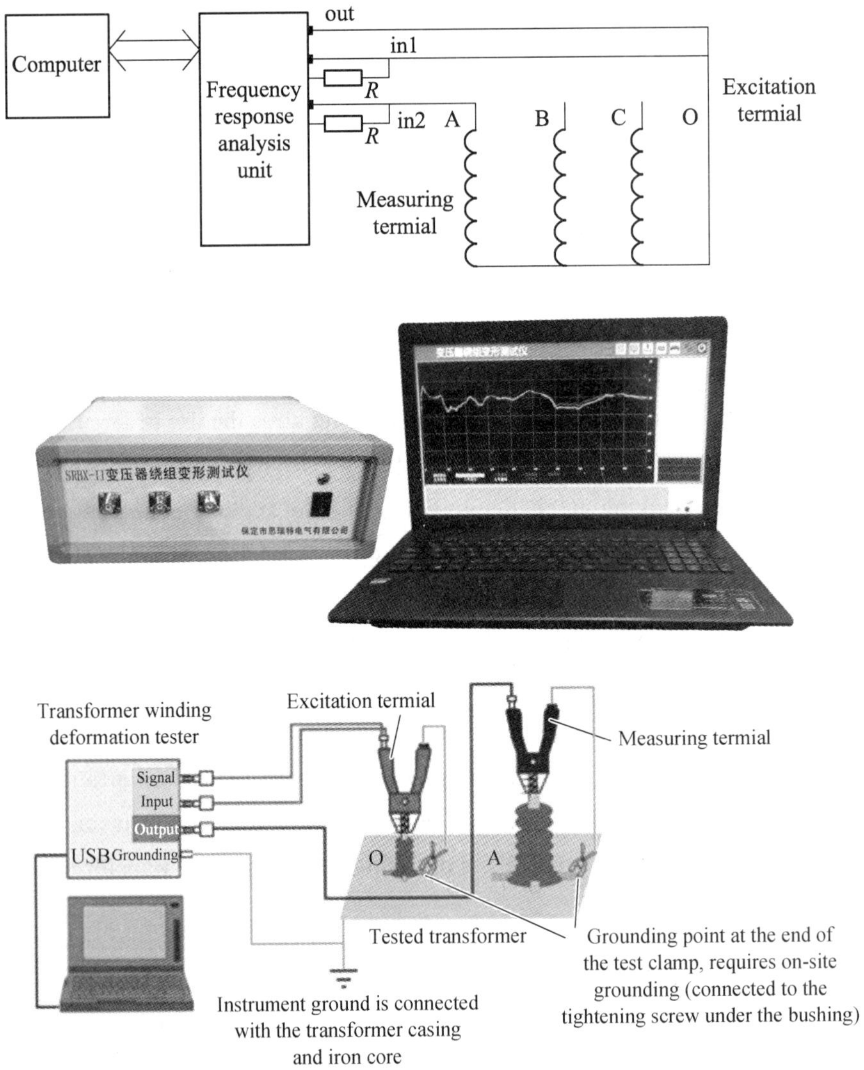

Figure 3-30 Wiring diagram for transformer winding deformation testing

The common wiring connection for transformer deformation testing is shown in Figure 3-31.

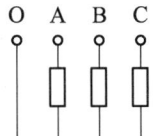
Input at O terminal, measurement at A terminal
Input at O terminal, measurement at B terminal
Input at O terminal, measurement at C terminal

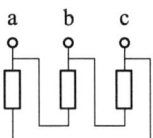
Input at a terminal, measurement at b terminal
Input at b terminal, measurement at c terminal
Input at c terminal, measurement at a terminal

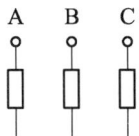
Input at A terminal, measurement at B terminal
Input at B terminal, measurement at C terminal
Input at C terminal, measurement at A terminal

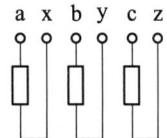
Input at a terminal, measurement at x terminal
Input at b terminal, measurement at y terminal
Input at c terminal, measurement at z terminal

Figure 3-31　Several common measurement wiring methods for transformer

Ⅱ. Test Steps

Select the excitation end (input teminal) and response end (measurement teminal) of the tested transformer, and use two bare copper wires to connect the "GND" grounding end of the input cable and the detection cable to the metal casing of the transformer oil tank together, ensuring reliable connection (contact resistance not greater than 1 Ω) with the outer shell. The grounding wire should be as short as possible and should not be tangled. It is recommended to connect it to the grounding copper bar position of the core grounding lead-out end. It is strictly prohibited to randomly wind it on the metal bolt of the oil tank's outer shell, otherwise, it will affect the measurement results.

Connect the input cable and detection cable to the corresponding sleeves at the selected excitation end and response end respectively using two wire clamps, and connect the Vs and V1 terminals of the input unit to the Vs and V1 ports of the test instrument respectively through a coaxial cable. Connect the V2 terminal of the detection unit to the V2 port of the test instrument correspondingly.

The two wiring clamps connect the input cable and the detection cable to the selected excitation end and the response end bushing end respectively. The Vs and V1 ends of the input unit are connected to the Vs and V1 ports of the tester through the coaxial cable, and the V2 end of the detection unit is connected to the V2 port of the tester.

If the software is installed on a laptop, connect the test instrument to the laptop using a dedicated serial cable. If the software is integrated into the instrument, simply power on the instrument and configure the settings within the instrument. After entering the parameters of the tested transformer nameplate values, frequency sweep range, display mode, etc. In the testing software, the test can be started. Refer to Figure 3-32 for details.

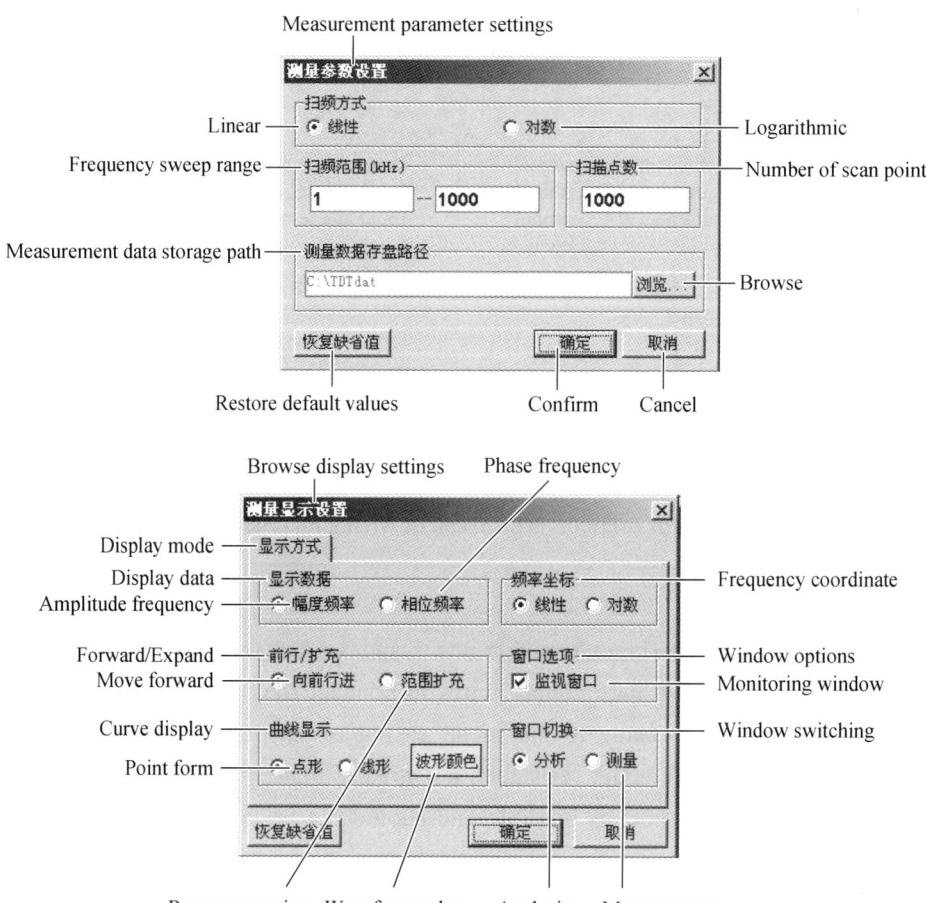

Figure 3-32 Deformation measurement data settings

After the data collection is complete, save the results and turn off the power of the laptop and the test instrument. The measurement is shown in Figure 3-33, where the horizontal axis represents the frequency sweep range and the vertical axis represents the response amplitude value.

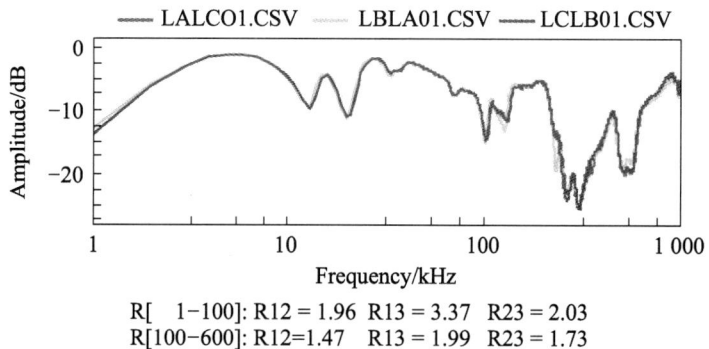

R[1–100]: R12 = 1.96　R13 = 3.37　R23 = 2.03
R[100–600]: R12=1.47　R13 = 1.99　R23 = 1.73

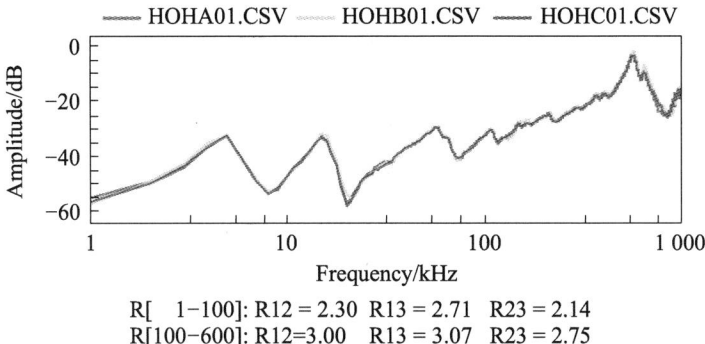

R[1–100]: R12 = 2.30 R13 = 2.71 R23 = 2.14
R[100–600]: R12=3.00 R13 = 3.07 R23 = 2.75

Figure 3-33 Transformer deformation measurement data

III. Criteria for Transformer Winding Deformation

1. Frequency Response Characteristics

The frequency response curve of the transformer winding exhibits rich resonance points in the medium-frequency and low-frequency range (10–500 kHz). These resonance points sensitively reflect the deformation of the winding, such as broken strands, bulges, distortions, and inter-disk displacement. The high-frequency range (above 500 kHz) reflects the displacement of the transformer winding. For the high-frequency portion of the frequency response curve of the winding at 110 kV and above, due to various influencing factors, sometimes it is difficult to ensure a good coincidence of this portion of the curve.

(1) Low-frequency Range

The normal transformer winding low-frequency band correlation coefficient R_{LF} requires to be greater than 2.0. In the amplitude-frequency response characteristic curve of the low-frequency range (1 kHz to 100 kHz), significant changes occur in the position of the peaks or valleys. This usually indicates a change in the inductance of the winding, which may be due to inter-turn or inter-disk short circuits. At lower frequencies, the capacitive reactance formed by the winding-to-ground capacitance and inter-disk capacitance is relatively high, while the inductive reactance is relatively low. If the inductance of the winding changes, it will cause significant movement in the position of the peaks or valleys in the low-frequency range of the frequency response characteristic curve. For the majority of transformers, the response characteristic curves of the three-phase windings in the low-frequency range should be very similar. If there are differences, the cause should be clearly identified.

(2) Mid-frequency Range

The normal transformer winding mid-frequency band correlation coefficient R_{MF} requires to be greater than 1.0. In the amplitude-frequency response characteristic curve of the mid-frequency range (100 kHz to 600 kHz), significant changes occur in the position of the peaks or valleys. This usually indicates local deformation phenomena such as twisting and bulging of the winding. The amplitude-frequency response characteristic curve in this frequency range has a larger number of peaks and valleys, which can sensitively reflect the changes in the distributed inductance and capacitance of the winding.

(3) High-frequency Range

The normal transformer winding high-frequency band correlation coefficient R_{HF} requires to be greater than 0.6. In the amplitude-frequency response characteristic curve of the high-frequency range (>600 kHz), significant changes occur in the position of the peaks or valleys. This usually indicates a change in the winding-to-ground capacitance, which may be due to overall displacement of the coil or lead displacement. At higher frequencies, the inductive reactance of the winding is larger, while the capacitive reactance is smaller. Since the inter-disk capacitance of the winding is much larger than the ground capacitance, the distribution of peaks and valleys is mainly influenced by the ground capacitance.

2. Deformation Comparison Method

(1) Longitudinal Comparison Method: It compares the amplitude-frequency response characteristics of the same transformer, same winding, and same tap switch position at different times to analyze the degree of winding deformation based on the changes in the amplitude-frequency response characteristics. This method has high detection sensitivity and accuracy, but it requires obtaining the original amplitude-frequency response characteristics of the transformer in advance.

(2) Horizontal Comparison Method: It compares the amplitude-frequency response characteristics of the three-phase windings of transformers with the same voltage level. If necessary, reference can be made to the amplitude-frequency response characteristics of the same model transformers manufactured by the same manufacturer during the same period to determine whether the winding deformation has occurred in the transformer. This method does not require the original amplitude-frequency response characteristics of the transformer and is more convenient for on-site application. However, it is necessary to exclude the possibility of similar deformation occurring in the three-phase windings of the transformer or the possibility of differences in the amplitude-frequency response characteristics of normal transformer three-phase windings themselves.

The amplitude-frequency response characteristic curve graph of a certain transformer's low-voltage winding before and after experiencing a sudden short-circuit current shock is shown in Figure 3-34. Comparing the post-shock amplitude-frequency response characteristic curve (LaLx02) with the pre-shock curve (LaLx01), it is evident that the frequency distribution position of some peaks and valleys have noticeably shifted to the right. This indicates that deformation has occurred in the transformer winding.

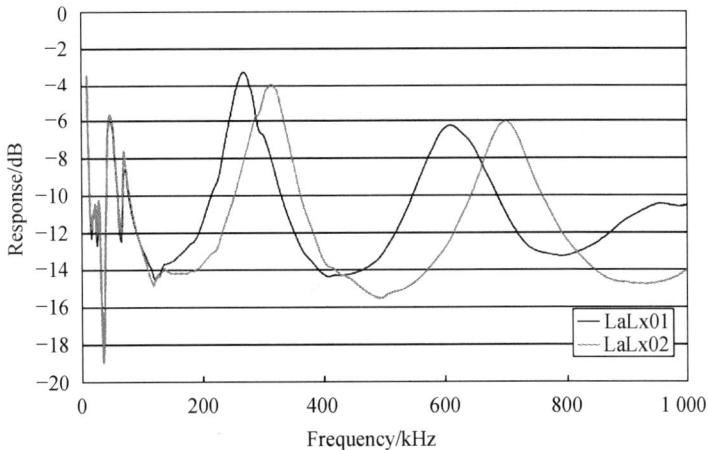

Figure 3-34 Amplitude-frequency response characteristic curve of a certain transformer's low-voltage winding before and after experiencing a sudden short-circuit current shock (longitudinal comparison method)

In Figure 3-35, the amplitude-frequency response characteristic curve of a three-phase transformer's low-voltage winding after experiencing a short-circuit current shock is shown. It is shown that compared with curves LcLa and LaLb, LbLc, there are differences in the frequency distribution position and the number of peaks and valleys in the amplitude-frequency response characteristic curve, indicating poor consistency among the three-phase windings of the transformer. However, the frequency response characteristics of the three-phase windings of another same-model transformer manufactured by the same manufacturer during the same period are more consistent (shown in Figure 3-36). Therefore, it can be concluded that deformation has occurred in the transformer winding after experiencing a sudden short-circuit current shock.

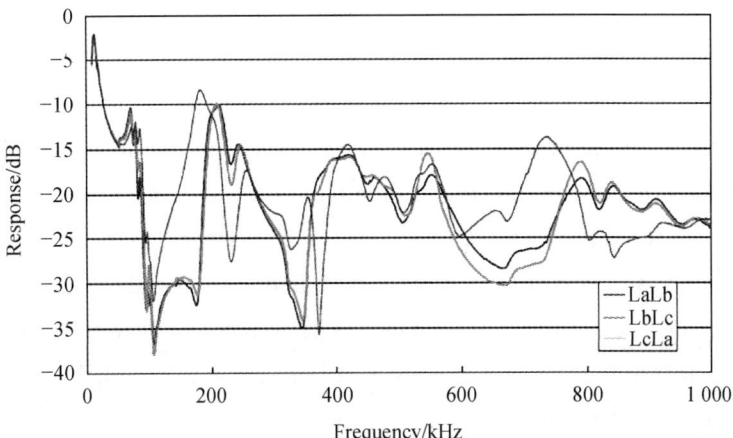

Figure 3-35 Amplitude-frequency response characteristic curve of a certain transformer's low-voltage winding after experiencing a sudden short-circuit

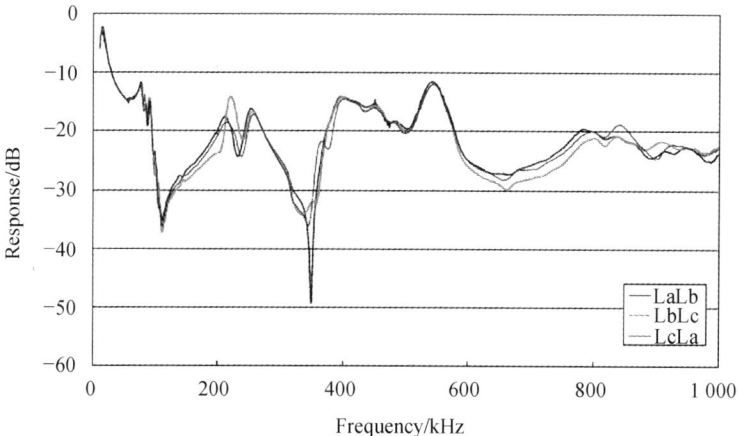

Figure 3-36 Amplitude-frequency response characteristic curve of
another same model transformer's low-voltage winding

3. Typical cases of transformer deformation curves

In the low-frequency range (several tens of kHz), the frequency response curves generally overlap well. Otherwise, poor test-wiring connections should be suspected first. Generally, the consistency of frequency response characteristics for the transformer up to 35 kV (including the substation transformer) may be poor, so the original data should be kept for comparison during handover acceptance.

The measured frequency response curves are generally between +20 dB and -80 dB. If they exceed this range, the test circuit should be checked for poor contact or breakage. When conducting a separate test on the angle-displaced windings, the three-phase frequency response characteristics may be inconsistent. Severe winding deformation can affect the frequency response characteristics of adjacent windings. Poor craftsmanship may lead to inconsistent frequency response characteristics of the transformer windings.

After detecting the deformation of transformer windings, longitudinal comparison and horizontal comparison are the main analysis methods, especially the longitudinal comparison. To judge the winding deformation and internal insulation of the main transformer based on the change in capacitance, it is not necessary to calculate the capacitance between individual windings and between windings-core and yoke for each pre-test. It is only necessary to compare the results of two tests. By observing the change in capacitance, a basic judgment can be made on the condition of the main transformer at the test site. Therefore, it is essential to establish and maintain accurate archival data of stable-state parameters for transformers.

Figure 3-37 shows the typical amplitude-frequency response characteristic curve of transformer winding deformation.

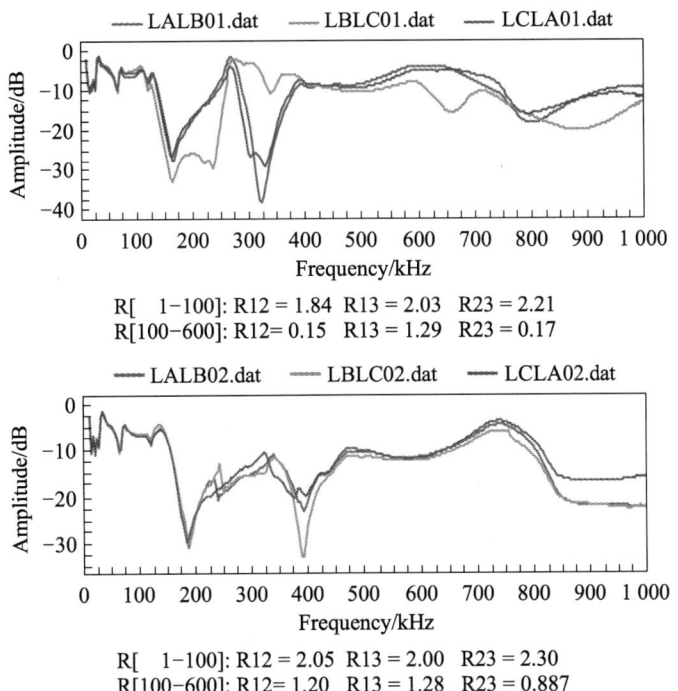

(a) Frequency response curves of SFPSZ7-120000/220 transformer with low voltage winding deformation (up figure) and after repair (down figure)

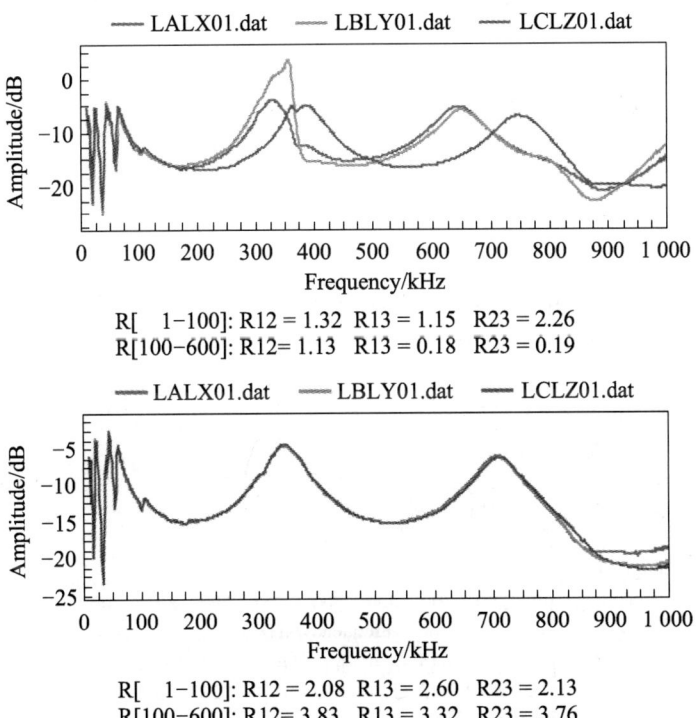

(b) Frequency response curves of SFPSZ7-150000/220 transformer with low voltage winding deformation (up figure) and after repair (down figure)

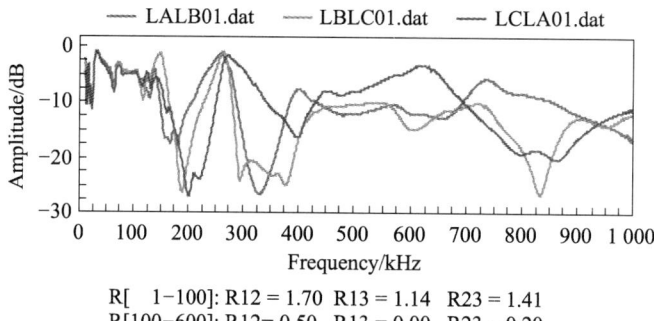

R[1–100]: R12 = 1.70 R13 = 1.14 R23 = 1.41
R[100–600]: R12 = 0.50 R13 = 0.00 R23 = 0.20

(c) Frequency response curve of SFPSZ7–150000/220 transformer low-voltage windings when deformed

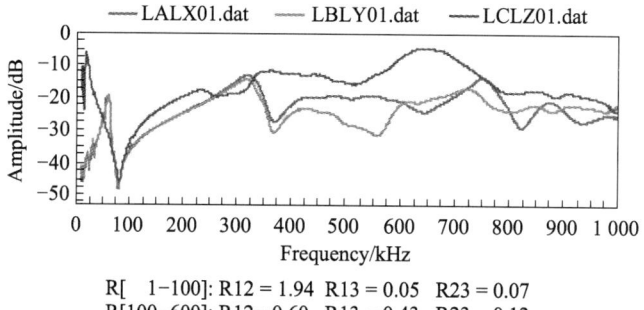

R[1–100]: R12 = 1.94 R13 = 0.05 R23 = 0.07
R[100–600]: R12 = 0.60 R13 = 0.43 R23 = 0.12

(d) Frequency response curve of SF9–31500/110 transformer low-voltage windings when deformed

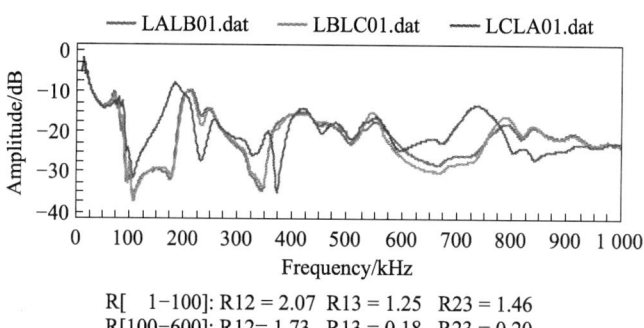

R[1–100]: R12 = 2.07 R13 = 1.25 R23 = 1.46
R[100–600]: R12 = 1.73 R13 = 0.18 R23 = 0.20

(e) Frequency response curve of SF7–6300/110 transformer low-voltage windings when deformed

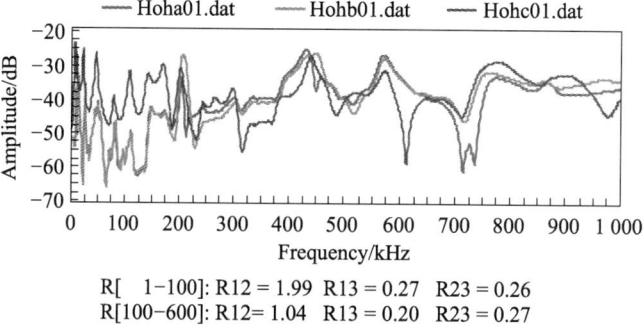

R[1–100]: R12 = 1.99 R13 = 0.27 R23 = 0.26
R[100–600]: R12 = 1.04 R13 = 0.20 R23 = 0.27

(f) Frequency response curve of SSPSO3–120000/220 transformer high-voltage windings when deformed

Figure 3-37   The typical amplitude-frequency response characteristic curve of transformer winding deformation

(I) Diagnostic steps for the absence of deformation in transformer windings with a capacity of 1.6 MVA or above

(1) If the interphase difference between the three-phase windings is greater than 3.5 dB, the test results should be compared with the original test results of the transformer. If there is a noticeable increase (greater than 3.5 dB), it can be determined that the winding has deformed.

(2) If there are no original test results available, the test results of transformers of the same model and period from the same manufacturer can be compared. If there is a significant increase (greater than 3.5 dB), it can be determined that the winding has deformed.

(3) If a comparison still cannot be made, it is necessary to request the manufacturer to explain the reason for the inconsistency in the interphase difference of the three-phase windings and make a judgment based on short-circuit and overcurrent conditions.

(4) If the interphase difference between the three-phase windings is less than 3.5 dB but the difference compared to the original test results of the transformer is greater than 3.5 dB, it indicates that the shared portion of the transformer winding has deformed or there has been a consistent deformation across all three phases.

(II) Diagnosis of Deformation Degree

The degree of deformation is determined based on the transformer winding deformation measurement results. The degree of deformation is divided into four types: normal, mild deformation, obvious deformation, and severe deformation. A normal degree of deformation indicates that the transformer is in its original state or has no significant deformation, and can continue to operate without the need for winding repair. Mild deformation refers to a situation where the transformer has obvious deformation but can still operate normally. It requires increased monitoring and appropriate maintenance scheduling, as any subsequent short circuits or impacts are likely to cause substantial damage to the transformer, necessitating winding repairs or replacement. obvious or severe deformation refers to a scenario where the transformer cannot continue operating due to deformation, and immediate handling is necessary.

The degree of transformer deformation and the corresponding diagnostic values are shown in Table 3-8.

Table 3-8 Transformer Deformation Degree and correlation coefficient, R

Transformer Deformation Degree	The correlation coefficient, R
Normal	$R_{LF} \geqslant 2.0$, $R_{MF} \geqslant 1.0$ or $R_{HF} \geqslant 0.6$
Mild deformation	$2.0 > R_{LF} \geqslant 1.0$ or $0.6 \leqslant R_{MF} < 1.0$
Obvious deformation	$1.0 > R_{LF} \geqslant 0.6$ or $R_{MF} < 0.6$
Severe deformation	$R_{LF} < 0.6$

Note: R_{LF} is the correlation coefficient within the low-frequency range (1kHz to 100 kHz); R_{MF} is the correlation coefficient within the mid-frequency range (100 kHz to 600 kHz); R_{HF} is the correlation coefficient within the high-frequency range (600 kHz to 1,000 kHz).

If the three-phase frequency response curves are consistent, the test data can be considered correct. If there are significant differences, the first step is to check whether the test wiring meets the specified requirements and whether the test cables are in good condition. After confirming that there are no issues, the test should be repeated. If the frequency response curve obtained from the retest is completely identical to the previous one, then the test data can be considered accurate.

By utilizing the historical test data of this transformer or the test data of another transformer of the same model and batch, a longitudinal comparative analysis can be conducted to make a relatively reliable diagnostic conclusion. If the test data of this transformer is essentially consistent with the historical data (or factory test data) without significant differences, it can be determined that the transformer does not exhibit obvious winding deformation, regardless of whether the frequency response curves between the three phases are consistent. This diagnostic method is reliable.

If there is no historical data available for this transformer, the test data of this transformer should be compared with the data of another transformer of the same model and batch. If they are basically consistent with no significant differences, it can be concluded that the transformer does not exhibit obvious winding deformation, regardless of whether the frequency response curves between the three phases are consistent.

In conducting longitudinal comparisons, it is important to consider the impact caused by differences in testing instruments, test wiring, and test conditions. As long as the frequency response curves of the two tests are essentially similar, or the patterns of differences between the three-phase windings do not show significant changes, it can generally be assumed that there is no obvious deformation in the windings.

Even if the preliminary analysis suggests the possibility of significant winding deformation, it is necessary to obtain the following information about the transformer and draw a comprehensive conclusion through further analysis:

(1) The manufacturing year or design model of the transformer. Based on experience, transformers produced between 1985 and 1998 have a lower ability to withstand transient short circuits and are more prone to winding deformation.

(2) Whether there have been cases of winding deformation reported for this particular model of transformer. Based on experience, transformers of the same model (especially from the same batch) often exhibit similar inherited defects.

(3) Whether the transformer has experienced external or nearby short-circuit impacts, whether the protective relays operated within the specified time, or whether the transformer was subjected to impacts during transportation or lifting.

(4) In addition, for the transformer produced by small factories or those that have undergone on-site repairs, the consistency of the frequency response characteristics among the three phases may be relatively poor. In such cases, it is generally advisable to relax the judgment criteria appropriately.

IV. Precautions

The detection of winding deformation in the transformer should be carried out before the DC test or after the winding has been fully discharged for at least 2 hours. Otherwise, it may affect the repeatability of the test data or even damage the testing instrument. Before the test, all leads connected to the transformer bushing terminals should be removed, and the removed leads should be kept as far away from the tested transformer bushing as possible. The nameplate and tap position of the tested transformer should be recorded. For the transformer where the bushing leads cannot be removed, the tap shielding end of the bushing can be used as the response terminal for testing, indicating that the test is conducted directly at the tap shielding end or externally connected to the transformer.

The frequency response characteristics of transformer windings are related to the tap switch position. It is recommended to measure at the maximum tap position or ensure that the tap switch is in the same position for each measurement. Due to the weak measurement signal, the excitation signal and response signal measurement terminals should be reliably connected to the transformer winding terminals to reduce contact resistance. The ground wire of the input unit and detection unit should be reliably connected to the transformer casing or oil tank, and there should not be a contact resistance greater than 1 Ω. The ground wire should be as short as possible and should not be wound.

It is usually recommended to connect it to the iron core grounding copper busbar located at the top of the transformer, and it is strictly prohibited to randomly wrap it around the bolts on the surface of the oil tank. After arranging the test cable, first, short-circuit the cable and check if the testing instrument and cable are intact.

When comparing the amplitude-frequency response characteristics of three-phase windings with the same voltage level, if there is a significant difference, the test cable and grounding leads should be checked, and the measurement should be repeated to ensure the repeatability of the measurement results for the same winding and eliminate the influence of factors such as measurement wiring. If the smoothness of the measured frequency response curve is poor (the curve has a lot of spikes), check whether the wire clamp is reliably connected to the bushing end, and whether the cable connector has poor contact or broken wires.

If the consistency of the frequency response characteristics of the three-phase windings is poor, check or change the connection position of the ground wire in the input cable and measurement cable, check whether it is reliably connected to the transformer casing, and repeat the measurement to ensure that the data curves obtained twice are completely consistent. Pay special attention to the fact that there is usually a contact resistance between the bolts on the transformer oil tank and the casing. Avoid using bolts for connections to prevent measurement wiring damage or incorrect test data caused by poor grounding lead connection to the casing.

The cable has good contact with the instrument and the tested transformer. For windings of transformers with a voltage level of 110 kV and above, the stray capacitance between the test leads and the bushing may affect the consistency of the high-frequency portion of the frequency response curve. Therefore, it is preferable to maintain consistency in consecutive tests or three-phase tests. If there is inconsistency in the frequency response characteristics of the three-phase windings during the test, the equipment should be checked and the test should be repeated until the results of the same phase are consistent. After the test is completed, check if the data files are properly saved, then exit the testing system and shut down in order.

Winding deformation testing is a new inspection method for the transformer. It can promptly and effectively detect winding deformation defects caused by short-circuit impacts on the transformer. Through timely maintenance and major repairs, major accidents can be avoided from occurring.

Module 4 Training

1. What are the types of transformer winding deformation, and how should they be handled?
___
___
___

2. Why is the transformer winding deformation test conducted before the transformer DC resistance test?
___
___
___

Worksheet 5 Transformer Oil Chromatographic Analysis

Module 1 Operating Worksheet: Oil Chromatographic Analysis

(Ⅰ) Test Name and Instrument	(Ⅱ) Test Objects
Transformer oil chromatographic analysis **Transformer oil chromatographic analysis instrument**	Oil-immersed power equipment
(Ⅲ) Test Purpose	(Ⅳ) Measurement Steps
The analysis of dissolved gases in oil by chromatography can detect the nature, degree, and location of latent faults in oil-immersed equipment at an early stage, thereby preventing accidents	(1) Take oil samples from the transformer body and on-load tap-changer sampling valves. (2) Turn on the power supply of the oil chromatographic analyzer. (3) Select the analysis method and set the instrument parameters. (4) Start measuring the sample. (5) Record and save the data
(Ⅴ) Precautions	(Ⅵ) Technology Standards
(1) Oil sampling should be done on dry and sunny days. (2) The oil storing container should be thoroughly cleaned and dried before use. (3) Oil samples should be taken from the drain valve at the bottom of the oil-immersed equipment. Wipe the valve clean, drain off the dirty oil, and take the sample after the oil is clean. (4) Minimize contact with the external air as much as possible to ensure there are no air bubbles in the oil sample. (5) After taking the oil sample, close the oil valve tightly to prevent oil leakage. (6) Check if the nitrogen, air, and hydrogen gas cylinders have sufficient pressure	(1) GB/T 17623–2017 "Gas Chromatographic Determination of Dissolved Gas Components in Insulating Oil." (2) DL/T 722–2014 "Guidelines for Analysis and Judgment of Dissolved Gases in Transformer Oil."

(Ⅶ) Result Judgment	(Ⅷ) Digital Resources
(1) Analyze the reasons and variations of the generated gases. (2) Determine the presence and types of faults, such as overheating, arc discharge, spark discharge, and partial discharge. (3) Assess the condition of the faults, including hotspot temperature, severity of the fault circuit, and development trends. (4) Propose corresponding remedial measures, such as whether operations can continue, technical safety measures and monitoring methods during operation, or whether hanging core detection is required. If enhanced monitoring is needed, the testing cycle for the next examination should be shortened	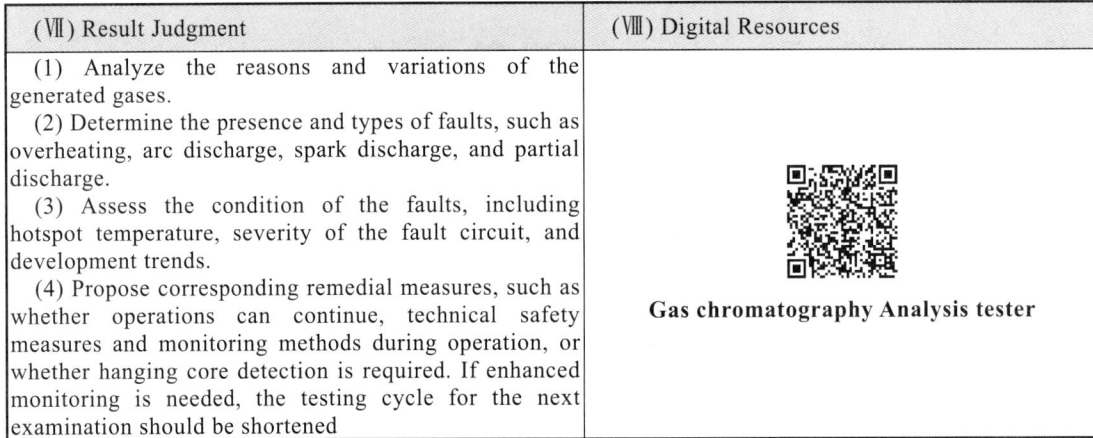 **Gas chromatography Analysis tester**

Module 2 Follow Me

Ⅰ. Gas Chromatographic Analysis of Transformer Oil

(Ⅰ) Significance of Transformer Oil Analysis

Gas chromatographic analysis is primarily used in the power system to detect dissolved gases in oil-immersed electrical equipment. Under normal circumstances, the insulating oil and organic insulation materials in oil-filled electrical equipment gradually age and decompose under the influence of heat and electricity, resulting in the generation of various low-molecular-weight hydrocarbons, carbon dioxide, carbon monoxide, and other gases. Most of these gases are dissolved in the oil.

When there is latent overheating or discharge fault in the equipment, the generation rate of these gases is accelerated. As the fault develops, the decomposed gases form bubbles in the oil, which continuously dissolve in the oil through convection and diffusion. By regularly analyzing the gases dissolved in the oil using gas chromatographic analysis during the equipment operation, potential internal faults in the equipment can be detected early, and the development of the fault can be monitored, allowing necessary measures to be taken in a timely manner.

(Ⅱ) Generation of Gases in Transformer Oil

The insulation of oil-immersed electrical equipment (such as transformer, reactor, current transformer, oil-immersed bushing, and oil-immersed cable) mainly consists of mineral insulating oil and organic insulation materials immersed in the oil (such as cable paper, insulating cardboard, etc.). Among them, mineral insulating oil, also known as transformer oil, is a distillation product of petroleum. Its main components are alkanes(C_nH_{2n+2}), cycloalkanes(C_nH_{2n}), aromatic unsaturated hydrocarbons(C_nH_{2n-2}), and other compounds. Organic insulation materials are mainly composed of cellulose ($C_6H_{10}O_5$).

Under normal operating conditions, both the oil and solid insulation gradually age and deteriorate, resulting in the decomposition and generation of small amounts of gas (primarily H_2, CH_4, C_2H_6, C_2H_4, C_2H_2, CO, and CO_2, etc.). When there is overheating, discharge faults, or moisture in the internal of the electrical equipment, the production of these gases increases rapidly. Most of these gases dissolve in the insulating oil, while a small portion rises to the surface of the oil. For example, in transformers, some gases escape from the oil and enter gas relays (gas detectors). The composition and content of various gases in the oil directly relate to the nature and severity of faults. Therefore, by regularly measuring the composition and content of gases dissolved in the oil during equipment operation, potential internal faults in oil-immersed electrical equipment can be detected early.

(III) Causes of characteristic gases

The various gas components in the oil can be obtained from oil samples taken from the transformer and analyzed by a gas chromatograph after degassing. Based on the content, characteristics, component ratios (such as the three-ratio), and gas generation rate of these gases, the internal faults of the transformer can be determined. In practical application, it is not appropriate to solely rely on the gas content in the oil as the sole criterion for determining the presence or absence of faults in the equipment. Instead, a comprehensive judgment should be made based on various possible factors. The gases and their causes during internal faults in the transformer are shown in Table 3-9.

Table 3-9 Gas and Cause during Internal Faults in the Transformer

Gas	Cause of Generation	Gas	Cause of Generation
H_2	Corona discharge, thermal decomposition of oil and solid insulation, moisture	CH_4	Thermal decomposition and discharge of oil and solid insulation
CO	Heating and decomposition of solid insulation	C_2H_6	Thermal decomposition and discharge of solid insulation
CO_2	Heating and decomposition of solid insulation	C_2H_4	Thermal decomposition and discharge of oil and solid insulation at high temperature hot spots
Hydrocarbon gas		C_2H_2	Intense arc discharge, thermal decomposition of oil and solid insulation

II. Gas Chromatographic Discrimination and Analysis of Power Transformer

Currently, in the fault diagnosis of the power transformer, it is often difficult to detect certain localized faults and thermal defects through electrical tests alone. However, the method of chemical detection through gas chromatographic analysis of gases in transformer oil is highly sensitive and effective in detecting certain latent faults inside the transformer and assessing their development at an early stage.

When there is an occurrence of overheating fault, discharge fault, or internal insulation moisture in the transformer, the content of these gases gradually increases. The gas components corresponding to the increased content due to these faults are shown in Table 3-10.

Table 3-10 Changes in Gas Components for Different Insulation Faults

Fault Type	Major Increased Gas Components	Minor Increased Gas Components	Fault Type	Major Increased Gas Components	Minor Increased Gas Components
Oil Overheating	CH_4、C_2H_4	H_2、C_2H_6	Arc in oil	H_2、C_2H_2	CH_4、C_2H_4、C_2H_6
Oil and Paper Overheating	CH_4、C_2H_4、CO、CO_2	H_2、C_2H_6	Arc in oil and paper	H_2、C_2H_2、CO、CO_2	CH_4、C_2H_4、C_2H_6
Partial Discharge in Oil and Paper	H_2、CH_4、C_2H_2、CO	C_2H_6、CO_2	Moisture or air bubbles in oil	H_2	
Spark discharge in oil	C_2H_2、H_2				

When conducting an internal fault diagnosis of the transformer based on chromatographic analysis, the following aspects should be included:

(1) Analyze the causes and changes of gas production.

(2) Determine whether there is a fault and the type of fault, such as overheating, arc discharge, spark discharge, and partial discharge.

(3) Assess the condition of the fault, such as hotspot temperature, severity of the fault circuit, and development trend.

(4) Propose corresponding treatment measures, such as whether to continue the operation, technical safety measures and monitoring methods during operation, or whether core lifting inspection is required. If enhanced monitoring is necessary, the testing cycle should be shortened.

III. The Relationship Between Changes in Characteristic Gases and Internal Faults of the Transformer

(I) Criteria for Judging Transformer Oil Faults

The "Regulations" stipulate the content of dissolved gases in the transformer. If any of the values exceed the specified standards, pay attention to identifying the cause of gas generation. Continuous monitoring should be conducted to assess the presence of faults, the severity of the faults, and their development trend. The standard for dissolved gas content in the transformer is shown in Table 3-11.

Table 3-11 Specified Values for Gas Content in Transformer Oil

Gas Component	Total hydrocarbons (methane, ethane, ethylene, acetylene)	Acetylene	Hydrogen
Content/($\mu L/L$)	150	5	150

Notes: The noted value of acetylene content in 500 kV transformer is 5 μL/L.

According to the "Regulations", when the total gas generation rate of hydrocarbon gases is greater than 0.25 mL/h (open type) and 0.5 mL/h (sealed type), or the relative gas generation rate is greater than 10%/min, it can be judged that there are abnormalities inside the transformer.

The main gases produced by the decomposition of fiber insulation materials in transformers at high temperatures are CO and CO_2, while hydrocarbons are present in small amounts. When oil-paper insulation is exposed to arcing, more acetylene gas will be generated. Due to the large variability in the measurement results of CO and CO_2 gases, there are currently no corresponding standards specified.

The "Regulations" provide criteria for determining the deterioration of gas content in transformer oil. These criteria can be used to determine whether the transformer oil has deteriorated, but they cannot determine the nature and condition of the faults.

(Ⅱ) Qualitative Analysis of Transformer Oil Fault

The use of characteristic gas analysis can help determine the causes of transformer faults. The dissolved gases in the oil can reflect the electrical and thermal decomposition nature of the surrounding oil and paper insulation caused by the fault. The characteristics of the gases vary depending on the type of fault, the energy involved, and the insulation materials affected. There is a close relationship between the unsaturation of hydrocarbon gases produced at the fault location and the energy density of the fault source. Characteristic gas analysis can provide a visual and convenient way to roughly analyze and determine the types of faults. The qualitative analysis of characteristic gas components and fault locations is shown in Table 3-12.

Table 3-12 Qualitative Analysis of Characteristic Gas Component and Fault Location

Fault type	Main Components	Characteristic Gas Description	Possible Fault Location
Partial Discharge	H_2、CH_4	Total hydrocarbon low, $H_2 > 100$ μL/L, CH_4 as main component of hydrocarbons	Winding partial discharge, local discharge between tap switch contact
Spark Discharge	H_2	Total hydrocarbon low, $C_2H_2 > 10$ μL/L, high H_2 content	Winding short circuit, poor contact of decomposition switch, poor insulation
Arc Discharge	H_2、C_2H_2	High total hydrocarbon, high C_2H_2 as main component of hydrocarbons, high H_2 content	Winding short circuit, decomposition switch flashover, arc short circuit
General Overheating	CH_4、C_2H_4	Total hydrocarbon low, $C_2H_2 < 5$ μL/L	Conductor overheating, decomposition switch failure
Severe Overheating	CH_4、C_2H_4	High total hydrocarbons, $C_2H_2 > 5$ μL/L but not a main component of total hydrocarbons, high H_2 content	Metal conductor overheating (temperature exceeds 1,000 °C)

When the H_2 content increases while other gas components remain unchanged, it may be due to water ingress or the chemical reaction between water and iron caused by bubbles, or the decomposition or corona effect of water or gas molecules under high electric field strength.

The acetylene content is a key indicator for distinguishing between overheating and discharge faults. However, in most overheating faults, especially when there are high-temperature hotspots, a small amount of acetylene can also be generated. For example, when the temperature exceeds 1,000 °C, a significant amount of acetylene may appear, which can be caused by both high-energy discharges and conductor overheating. Acetylene can also occur when tap switches overheat. Low-energy partial discharges do not generate acetylene or only produce a minimal amount.

Under arc action, gases are generated from the decomposition of transformer oil and solid insulation, as shown in Table 3-13.

Table 3-13 Gases generated from decomposition of transformer oil and solid insulation under arc action (volume %)

gas	H_2	C_2H_2	CH_4	C_2H_4	CO	CO_2	O_2	N_2
Transformer oil	57–74	14–24	0–3	0–1	0–1	0–3	1–3	2–12
Oil-immersed Paperboard	40–58	14–21	1–10	1–11	13–24	1–2	2–3	4–7
Oil-phenolic Resin	41–58	4–11	2–9	0–3	24–35	0–2	1–3	2–6

(Ⅲ) Three-Ratio Method for Transformer Fault Diagnosis

The three-ratio method is used to determine the nature of transformer faults by comparing the ratios of three pairs of gases among five gases. Different codes are used to represent different three ratio values and different ratio ranges. The coding rules for the three-ratio method are shown in Table 3-14, and the diagnostic criteria for fault nature using the three-ratio method are shown in Table 3-15.

Table 3-14 Coding rules for the three-ratio method

Ratio of Characteristic Gas	Ratio Range Coding			Explanation
	C_2H_2/C_2H_4	CH_4/H_2	C_2H_4/C_2H_6	
<0.1	0	1	0	When C_2H_2/C_2H_4=1–3, the code is 1. When CH_4/H_2=1–3, the code is 2. When C_2H_4/C_2H_6=1–3, the code is 1
0.1–1	1	0	0	
1–3	1	2	1	
>3	2	2	2	

Table 3-15 Three-ratio method for determining fault nature

ID	Fault nature	Ratio Range Coding			Typical Examples
		C_2H_2/C_2H_4	CH_4/H_2	C_2H_4/C_2H_6	
0	No fault	0	0	0	Normal aging
1	Partial Discharges with Low Energy Density	0	1	0	Discharge in gas-filled voids, which are caused by incomplete impregnation, gas oversaturation, air entrapment or high humidity, etc.

Continued

ID	Fault nature	Ratio Range Coding			Typical Examples
		C_2H_2/C_2H_4	CH_4/H_2	C_2H_4/C_2H_6	
2	Partial Discharges with High Energy Density	1	1	0	Same as above, but discharge marks or perforation of solid insulation have been caused
3	Low Energy Discharge	1→2	0	1→2	Continuous spark discharge between poorly connected points at different potentials or between suspended potential bodies. Oil breakdown between solid materials
4	High Energy Discharge	1	0	2	Discharge with power frequency current. Arc breakdown of oil between coils, windings, or between coils and ground. Switching off the current with selector switches on load tap switches
5	Thermal Failure below 150 °C	0	0	1	Usually, it is the overheating of insulating conductors
6	Thermal Failure in the low temperature range of 150–300 °C	0	2	0	Due to the local overheating of the iron core caused by flux concentration, the hot spot temperature increases sequentially in the following cases: small hot spots in the iron core, short circuit in the iron core, and copper overheating caused by eddy current, poor joint or contact (forming coke), iron core and shell circulation
7	Overheating in the Moderate Temperature Range of 300–700 °C	0	2	1	
8	High-temperature Fault above 700 °C	0	2	2	

When there are high-temperature overheating and discharge faults inside the transformer, in most cases, $C_2H_2/C_2H_4 > 3$. Therefore, the remaining two items in the three-ratio method can be used to form a Cartesian coordinate system, with CH_4/H_2 as the ordinate and C_2H_2/C_2H_6 as the abscissa, forming a T (overheating) D (discharge) analysis judgment diagram as shown in Figure 3-38.

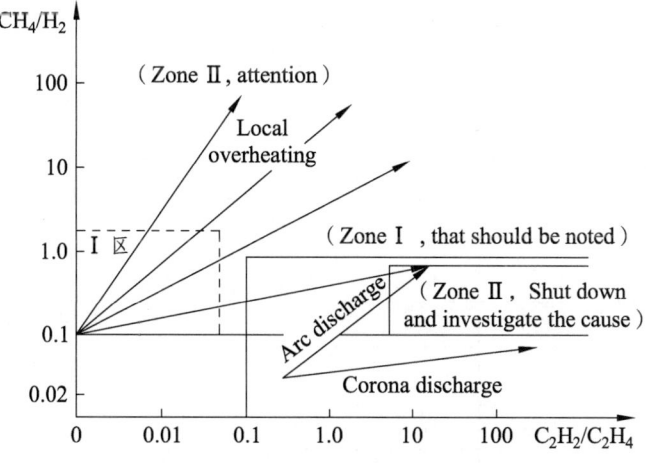

Figure 3-38　T-D analysis judgment diagram

The TD diagram method can distinguish whether the transformer is overheating or discharge fault, and divide the local overheating, corona discharge, and arc discharge areas according to their ratios. This method can quickly and correctly judge the nature of the fault, and play a monitoring role. Generally, internal faults of the transformer, except for discharge faults with suspended potentials, mostly start with overheating and develop towards the overheating area Ⅱ or discharge area Ⅱ. They end up with direct damage caused by overheating or discharge faults. The discharge area Ⅱ is a major hidden danger that needs to be strictly monitored and dealt with early. When the CH_4/H_2 ratio approaches 3, it may cause light gas action of the transformer and send a signal.

The ratio method based on dissolved gas analysis (DGA) in oil can be used to determine transformer faults. The dissolved gases in the oil, including H_2, CH_4, C_2H_6, C_2H_4, and C_2H_2, are used as characteristic data for the transformer's condition. The main fault modes of the power transformer during operation include high-temperature overheating, medium-temperature overheating, partial discharge, and arc discharge. For example, when the transformer experiences a high-temperature overheating fault (temperature > 700 °C), ethylene is the main characteristic gas, followed by methane, and their sum usually accounts for more than 80% of the total hydrocarbons. In the case of high-energy discharge, acetylene and hydrogen are the main fault characteristic gases, followed by ethylene and methane. Acetylene generally accounts for 20% to 70% of the total hydrocarbon content, and hydrogen accounts for 30% to 90% of the total hydrogen content. Generally, the ethylene content is higher than the methane content.

Based on the statistical analysis of a large number of detected faulty transformers, the transformer conditions are classified into nine categories: normal operating sequence, low-energy discharge fault sequence, high-energy discharge fault sequence, medium-temperature overheating fault sequence, high-temperature overheating fault sequence, screen dendritic discharge sequence, interturn and interlayer fault sequence, tap switch fault sequence, and core grounding fault sequence. Each fault sequence is described by a standard fault mode. The standard fault modes for transformer fault diagnosis are shown in Table 3-16.

Table 3-16 Nine transformer standard fault modes

Failure mode	H_2%	CH_4%	C_2H_6%	C_2H_4%	C_2H_2%	Note
X1	46.1	21.5	61.5	15.8	1.2	Normalcy
X2	58.0	44.9	11.0	20.6	23.5	Low-energy discharge
X3	43.7	30.2	3.7	46.6	19.4	High-energy discharge
X4	15.3	26.2	21.0	52.8	0	Medium-temperature overheating
X5	11.3	24.6	12.7	59.9	2.8	High-temperature overheating
X6	58.6	30.5	4.9	26.2	38.4	Screen dendritic discharge
X7	28.8	28.2	3.9	34.4	33.4	Transformer inter-turn and inter-layer faults
X8	13.6	21.3	10.8	58.1	9.5	Tap switch failure
X9	11.2	30.8	11.6	56.2	1.4	Core ground fault

IV. Transformer Online Oil Monitoring System

The transformer online oil monitoring system is applicable to the transformer and other power equipment, as shown in Figure 3-39.

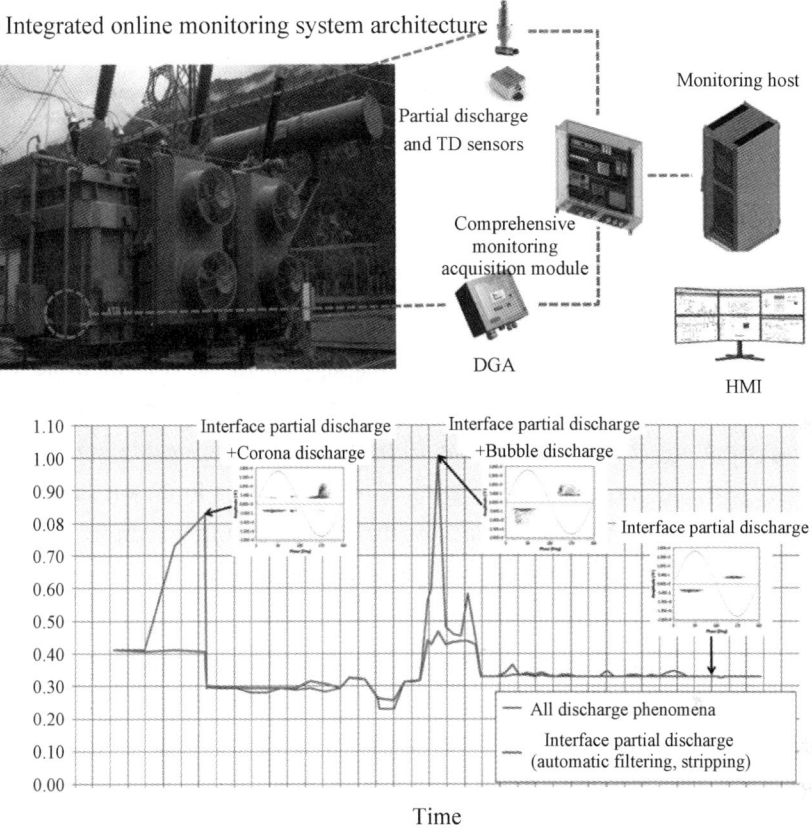

Figure 3-39 Structure of transformer oil comprehensive online monitoring system

The monitoring system should have the capability to distinguish different types of discharges and issue alarms for dangerous partial discharge phenomena, rather than discharge signals.

Module 3 Workshop

I. Oil Sampling

(1) The sampling container should be rinsed with distilled water or deionized water, followed by cleaning with isopropyl alcohol or high-purity ethanol that has been treated for cleanliness, and then dried for later use.

(2) The sampling valve for the transformer's operating oil should be cleaned with isopropyl alcohol or high-purity ethanol and allowed to air dry or wiped clean with Class A gauze. To avoid mixing residual oil from the valve body into the sample during sampling, 1–2 L of transformer oil

should be allowed to flow out naturally from the valve first, and the valve opening should be kept consistent during flushing and sampling.

(3) The sampling equipment should be placed properly to prevent collision between glassware during transportation and resulting in breakage.

II. Start-up steps

(1) Check if there is sufficient pressure in the nitrogen, air, and hydrogen gas cylinders. If the total pressure is below 2 MPa, replace the cylinders. Check the pressure gauge on the low-pressure side of the pressure-reducing valve, which is generally around 0.3 MPa. Check for leaks in the gas system, especially hydrogen. If any leaks are found, immediately locate the leaking point and take corresponding measures. Confirm the resolution before opening the gas cylinders.

(2) Needle cleaning: Before sampling, the glass needle should be thoroughly cleaned with transformer oil from the equipment. Follow the principle of small amounts multiple times, with each flushing using no less than 100 mL of oil. The number of flushes should be no less than 3 times.

(3) During sampling, ensure a tight connection between the needle and the rubber tube to minimize contact with the outside air. This ensures no air bubbles in the oil sample. Attach a label corresponding to the equipment name on each well-taken oil sample needle.

(4) Open the valves of the hydrogen, nitrogen, and air cylinders. Set the pressure of hydrogen gas to 0.2 MPa, and the pressure of nitrogen and air to around 0.3 MPa.

(5) Turn on the gas chromatograph and ensure that the meter on the left side of the gas chromatograph displays the correct reading. Set the test conditions and temperature correctly. If the gas chromatograph has not been used for a long time (more than six months), set the aging temperature (for how long should be aged), as shown in Table 3-17, Table 3-18 and Table 3-19.

Table 3-17　Normal readings on the meter of the gas chromatograph

Carrier gas I : 0.21	Carrier gas II : 0.06
Hydrogen I : 0.08	Hydrogen II : 0.11
Air I : 0.03	Air II : 0.03

Table 3-18　Test condition setting temperature

Column chamber: 60 °C	Thermal conductivity: 90 °C
Hydrogen flame: 120 °C	Conversion: 360 °C
Vaporization: 70 °C	

Table 3-19　Aging Temperature

Column chamber: 150 °C	Thermal conductivity: 120 °C (minimum)
Hydrogen flame: 220 °C	Conversion: 390 °C (important)
Vaporization: 220 °C	

(6) After confirming the readings on the meter and setting the temperature, press the run button, and the heating lamp will illuminate. Once the steady temperature lamp is illuminated, press the ignition button, and two "click" sounds will be heard. Use a small wrench to confirm that the flame has been ignited. Then open the bridge flow to ensure that the bridge flow lamp is on. Next, turn on the computer and then turn on the workstation. Confirm that the parameters of the gas chromatograph are correct, as shown in Table 3-20 and Table 3-21.

Table 3-20 Detection parameters

Sensitivity Ⅰ: 4	Sensitivity Ⅱ: 4
Attenuation: 2	Bridge current: 90

Table 3-21 Time parameters

T_1: 0:20	t_1 (follow changes automatically)
T_2: 7:40	t_2 (follow changes automatically)

Ⅲ. Setting Method for Standard Sample

Open the software and double-click on the analysis method. Set the method file name, device number, select shaking method, 220 kV and below, and device status (running state). Set the stop time to 8 minutes. Choose the printing options according to your needs, such as printing chromatographic analysis, printing analysis result data, and printing judgment analysis results. Click "Finish" to complete the method setup.

Select the analysis method and confirm the use of this analysis method. Click on "Create New Calibration Curve". Double-click on the component name and select the component name (in the order of hydrogen, carbon monoxide, carbon dioxide, methane, ethylene, ethane, acetylene). Enter the standard gas concentration for each component and confirm.

Select "Standard Sample" in the sample type, and set the standard sample number as Concentration 1 (switching between standard sample and test sample to achieve the corresponding effect described above). After that, inject a shot of the standard sample. Once the curve appears, assign a file name to it and confirm. Click on "Calibration Curve" to check for calibration factors (there should be). The red cross on the calibration curve should change to a checkmark.

Ⅳ. Set the alignment

After building the analysis method, inject a shot of standard gas and align the spectrum after it appears. Click on the system options to confirm that dual channels are selected for sampling and that the merged spectrum calculation mode is used for diagnosis. Click on file settings and select "Automatically merge spectra after sampling is completed" and "Merge channel B into channel A". With the alignment set, the icon in front of the calibration curve should change to a checkmark. Now, inject the gas sample for diagnosis.

The shutdown sequence for the instrument is as follows: Turn off the computer, turn off the workstation, turn off the chromatograph, and then turn off the gas. When shutting down the chromatograph, press the stop button first and wait for the conversion temperature to drop below 100 °C before turning off the power to the chromatograph. Note that the power on/off sequence must not be changed, and the temperature must be lowered to below 100 °C before turning off the power to the chromatograph. Do not touch anything related to the chromatograph while it is running, and do not take nitrogen gas or other substances during this period. If the gas from the tested sample contains hydrogen, methane, ethylene, ethane, and acetylene, the condition of the oil can be determined using the three-ratio method.

When sampling for sample testing, use a syringe to take 40 mL of the tested oil and plug the needle part of the syringe with a rubber cap, trying to avoid any gas in the oil. Use a nitrogen gas syringe (which needs to be cleaned) to take 5 mL of nitrogen gas and inject it into the oil to be tested. Place the tested oil in the oscillator steadily and press the confirm button to select the oscillation mode of 20 minutes of oscillation and 10 minutes of static state.

The oscillator begins to heat up, and when it reaches a constant temperature of 50 °C, it starts to vibrate and automatically alarms when the vibration is complete. After the oscillation is complete, use a syringe to take out the gas released from the oil, and record the amount of degassing, which is generally around 3 mL. Before testing, click on the sample icon, set the oil sample volume to 40 mL, and fill in the recorded degassing amount. Then, use a sampling needle to take 1 mL of the released gas and inject it into the inlet. After the analysis is complete, print out the results and finish.

Module 4 Training

What is the significance of transformer oil chromatographic analysis?

_____
_____
_____

Worksheet 6 Insulating Oil Dielectric Strength Test

Module 1 Operating Worksheet: Insulating Oil Dielectric Strength Test

(Ⅰ) Test Name and Instrument	(Ⅱ) Test Objects
Insulating oil dielectric strength test. **insulating oil dielectric strength test instrument**	Used for high-voltage electrical equipment such as transformer, oil circuit breaker, oil-filled cable, power capacitor, and oil bushing
(Ⅲ) Test Purpose	(Ⅳ) Measurement Step
To check whether the insulation performance of the insulating oil has decreased, whether it has aged, and the degree of contamination by water and other suspended substances, conducting a breakdown test on the insulating oil is the main method to assess its quality and performance	(1) Take an oil sample. (2) Clean the oil cup. (3) Place the oil cup between the high-voltage electrodes of the instrument. (4) Inject the oil sample into the oil cup. (5) Set the instrument parameters. (6) Press the run button to start the high-voltage test. (7) When the measurement is complete, record the results
(Ⅴ) Precautions	(Ⅵ) Technical Standards
(1) Clean the oil cup strictly according to the requirements. (2) Ensure that the instrument is well grounded. (3) Cover the test chamber with a high-voltage hood before powering on. (4) During the test, it is strictly forbidden to open or touch the high-voltage hood	Refer to national standards GB/T 507–2002 "Method for Determining Dielectric Strength of Insulating Oil", GB/T 4756–2015 "Method for Sampling of Petroleum and Liquid Petroleum Products (Manual Method)", and GB 2536–2011 "Transformer Oil".
(Ⅶ) Result Judgment	(Ⅷ) Digital Resources
(1) The first spark discharge voltage may be extremely low because of deviation and lower values. An average of 2–6 tests can be adopted. (2) The spark discharge voltage gradually increases because of moisture absorption capability due to increased oil humidity after spark discharge. (3) The six spark discharge voltages gradually decrease because of free-charged particles, bubbles, and carbon debris increasing, destroying the insulation performance of the oil. (4) The spark discharge voltage is low at both ends and high in the middle, which is a normal phenomenon	**Insulating Oil Dielectric Strength Test**

Module 2 Follow Me

Ⅰ. Principle of Dielectric Strength Testing for Insulating Oil

Dielectric strength testing for insulating oil is a commonly used method to assess the insulation performance of the oil, and it is an important means to evaluate the insulation property of insulating oil as a type of insulation material. The insulation performance of insulating oil is crucial for the safe operation of electrical equipment. The dielectric strength test for insulating oil determines its ability to withstand voltage under high electric fields, thereby determining its suitability for use in high-voltage equipment.

The principle of dielectric strength testing for insulating oil is mainly based on the effect of an electric field. During the test, a high voltage is applied to the insulating oil to generate a strong electric field. The insulation performance of the insulating oil determines its ability to withstand voltage under an electric field, known as the dielectric strength. During the test, two electrodes are inserted into the insulating oil, serving as positive and negative electrodes respectively, and a high voltage is applied. The electric field formed between the electrodes will conduct a breakdown test on the insulating oil to evaluate its insulation capability.

There are several key points to note during the process of dielectric strength testing for insulating oil. First, it is important to ensure good contact between the electrodes and the insulating oil, and the distance between the electrodes should meet the testing requirements. Second, it is important to ensure the uniform distribution of the electric field between the electrodes and the insulating oil, to avoid localized high electric field intensity. Additionally, the effect of temperature should be considered during the test, as there is a certain relationship between the dielectric strength of the insulating oil and temperature. Figure 3-40 shows an automatic insulating oil dielectric strength tester.

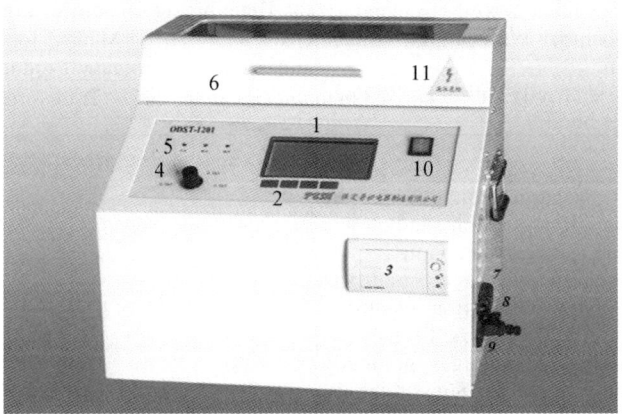

1–LCD; 2–Function keys; 3–Printer; 4–Boosting rate switch; 5–Indicator light; 6–Oil cup compartment cover;
7–Temperature and humidity sensors; 8–Ground line column; 9–Power socket;
10–Power switch; 11–High voltage safety sign.

Figure 3-40 Automatic insulating oil dielectric strength tester

Ⅱ. Overview of Insulating Oil

Insulating oil is widely used in high-voltage electrical equipment such as transformers, instrument transformers, switchgear, oil-filled cables, power capacitors, and bushings. During operation, the performance of insulating oil gradually deteriorates due to the effects of oxygen, high temperature, high humidity, sunlight, strong electric fields, and impurities, which prevent it from fully functioning as an insulator.

In order to confirm whether the insulation performance of insulating oil meets the requirements, regular tests on its relevant properties must be conducted. According to the national standard GB 507–2002 "Method for Testing Dielectric Strength of Insulating Oil," the dielectric strength of insulating oil refers to the breakdown voltage, breakdown strength, and voltage withstand capability of the oil. It is an important quality indicator that represents the ability of insulating oil to withstand voltage. The dielectric strength test gradually increases the voltage applied to the insulating oil under specified test conditions until it breaks down and loses its insulation properties. It is expressed as the average breakdown voltage divided by the electrode distance (kV/cm). A higher dielectric strength indicates better insulation performance.

Transformer oil is a fractional distillation product of petroleum, mainly composed of compounds such as alkanes, cycloalkanes, saturated hydrocarbons, and aromatic unsaturated hydrocarbons. It is a mixture of liquid natural hydrocarbons obtained from the lubricating oil fraction in petroleum through acid-base refining, which is pure, stable, has low viscosity, good insulation, and cooling properties. It is a light and yellow transparent liquid with a relative density of 0.895 and a freezing point below -45 °C. Transformer oil serves as an insulating and cooling lubricating medium in the oil-immersed transformer. The service life of transformer oil can usually reach 10–15 years.

(Ⅰ) Requirements for insulating oil

(1) It should have a high dielectric strength to adapt to different operating voltages.

(2) It should have a low viscosity to meet the needs of circulation convection and heat transfer.

(3) It should have a high flash point temperature to meet fire safety requirements.

(4) It should have sufficient low-temperature performance to withstand potential low-temperature environments, and good oxidation resistance to ensure a long service life of the oil.

(Ⅱ) Electrical, physical and chemical properties of transformer oil

Judging the quality of oil from its appearance, the color of new oil is generally light and yellow, but it turns dark red after oxidation. If the color of oil becomes dark quickly during operation, it indicates that the oil quality has deteriorated. Normal transformer oil is colorless and odorless. If there is a different odor, it means that the oil quality has deteriorated (as shown in Figure 3-41). For example, a burnt smell indicates that the oil has overheated when it dries out, an acidic smell indicates that the oil has seriously aged, and an acetylene smell indicates that arcing has occurred within the oil. New oil has good transparency and is transparent in a glass bottle with blue-violet fluorescence. If it loses its fluorescence and transparency, it indicates the presence of water, mechanical impurities, and free carbon.

Figure 3-41 Transformer oil samples

(Ⅲ) Electrical, Physical, and Chemical Properties Testing of Transformer Oil

The breakdown voltage of transformer oil is closely related to the cleanliness of new oil and the deterioration condition of operating oil. The electrical, physical, and chemical properties testing of transformer oil are shown in Table 3-22.

Table 3-22 Electrical, Physical, and Chemical Properties Testing of Transformer Oil

ID	Characteristic Indicators	Normal Oil Quality Indicator	Test Significance
1	Breakdown Voltage	The oil cup spacing is 2.5 mm. (1) $V_{rated} \leqslant 35$ kV: $V_{Break} > 35$ kV (New oil); $V_{Break} > 30$ kV (operating oil). (2) $V_{rated} = 66-220$ kV: $V_{Break} > 40$ kV (New oil); $V_{Break} > 35$ kV (operating oil). (3) $V_{rated} = 330$ kV: $V_{Break} > 50$ kV (New oil); $V_{Break} > 45$ kV (operating oil). (4) $V_{rated} = 500$ kV: $V_{Break} > 60$ kV (New oil); $V_{Break} > 50$ kV (operating oil)	Test the performance of transformer oil and detect impurities
2	Dielectric Loss	Not exceeding 0.005 (90 °C). (1) $V_{rated} \leqslant 330$ kV: $\tan\delta < 0.01$ (New oil); $\tan\delta < 0.04$ (operating oil). (2) $V_{rated} \leqslant 500$ kV: $\tan\delta < 0.007$ (New oil); $\tan\delta < 0.02$ (operating oil)	Detect the degree of transformer deterioration
3	Oil color	Transparent, without suspended particles and mechanical impurities	Detect oil aging
4	Acidity value	New oil not exceeding 0.03 mg KOH/g, operating oil not exceeding 0.1 mg KOH/g	The magnitude of acid value reflects the refining depth and oxidation degree of the oil

Continued

ID	Characteristic Indicators	Normal Oil Quality Indicator	Test Significance
5	Interfacial tension	At 25 °C. New oil not less than 35 mN/m, operating oil not less than 19 mN/m	Reflecting the degradation products of insulating oil and the soluble polar impurities produced from solid insulating materials
6	Moisture content	(1) $V_{rated} \leqslant 110$ kV: MC>20 μL/L (New oil), MC>35 μL/L (operating oil). (2) $V_{rated}=220$ kV: MC>15 μL/L (New oil), MC>25 μL/L (operating oil). (3) $V_{rated} \geqslant 330$ kV: MC>10 μL/L (New oil), MC>15 μL/L (operating oil)	Test the electrical performance and physical and chemical properties of the insulating oil
7	Flash point	New oil numbers 10 and 25 are not lower than 140 °C (closed cup), new oil number 45 is not lower than 135 °C (closed cup), and operating oil is not lower than 10 °C compared to the determined value of new oil	Due to internal faults in the equipment which cause the decomposition of insulating oil, volatile and flammable low-molecular hydrocarbons may be produced, leading to a decrease in flash point. Flash point testing reflects the ease of ignition of insulating oil and the amount of light distillates in the oil, reflecting the protective performance of the oil at high temperatures
8	Water-soluble acid	New oil pH > 5.4. Operating oil pH $\geqslant 4.2$	When transformer oil oxidizes, organic and inorganic acids are produced. When the oil contains water, water-soluble acids have good solubility in water, corroding solid insulating materials and metals, and damaging equipment insulation. New oil does not contain acidic substances and has low acid value. Test the pH value to determine the change in the content of water-soluble acids in the insulating oil

Ⅲ. Performance of Transformer Oil

Transformer oil serves as a medium within electrical equipment and must possess good electrical performance in order to fully function as insulation and cooling. The main electrical performances of new oil include insulation strength and dielectric dissipation factor.

(Ⅰ) Insulation Strength

Insulation strength refers to the dielectric strength or breakdown voltage of transformer oil. It measures the ability of the oil to withstand voltage within electrical equipment without being damaged. This is one of the main methods to assess the quality of transformer oil. The breakdown process of insulating oil is related to its purity. Pure insulating oil has high breakdown strength (up to 106 kV/cm), and its breakdown process is mainly caused by electrical breakdown. The oil in equipment often contains various impurities, such as gases, moisture, solid particles, and polymers generated from oil aging.

(Ⅱ) Dielectric loss factor

The dielectric loss factor primarily reflects the power loss caused by leakage current in the oil. The magnitude of the dielectric loss factor is sensitive in determining the degree of degradation and contamination of transformer oil. The dielectric loss factor can only indicate whether the oil contains pollutants and polar impurities, but it cannot determine the specific type of impurities present in the oil. However, when the oil deteriorates due to oxidation or overheating, or when it is mixed with other impurities, the dielectric loss factor increases as the content of impurities or colloidal substances charged in the oil increases, even up to more than 10%.

Dielectric loss tests are mainly used to determine whether the oil is dirty or degraded. They can only determine the presence of polar substances in the oil, but not the specific type of polar substances. In the case of further oxidation, the oil's ability to dissolve water may increase, and therefore the presence of sludge in this type of oil may not be reflected in the dielectric loss factor. If the dielectric loss exceeds 0.7%, an inspection is required. If the dielectric loss at 100 °C is 7–10 times higher than that at 25 °C, it indicates that the oil is dirty rather than containing moisture.

(Ⅲ) Physical and Chemical Property Tests

(1) Appearance and Color of Oil

Good oil should be clean and transparent. If it is unclear, it may contain water, carbon particles, or sludge. If carbon particles are found, there may be arcing or partial discharge phenomena inside the transformer, and it is necessary to perform chromatographic analysis on the oil. If there is a significant change in the color of the oil, pay attention to whether the oil is aging accelerated or the monitoring of the operating temperature of the oil.

(2) Acid value

Acid value refers to the acidic components contained in 1 g of test oil, expressed as mg KOH/g. It is an important indicator for testing transformer oil and reflects the degree of oil refinement and oxidation. The increase in acid value is a sign of initial oil deterioration, and the presence of acidic substances will inevitably generate sludge. If there is moisture in the oil, it can cause rusting of metals such as iron and also damage paper insulation.

(3) Interfacial tension test

The interfacial tension test is quite sensitive to degradation products in the oil and soluble impurities generated from solid insulation materials. The greater the content of oxidation products in the oil, the lower the interfacial tension. If the interfacial tension value in the oil is between 27–30 mN/m, it indicates a tendency for sludge formation. If the tension value drops 8 mN/m, it indicates severe oil aging and should be replaced.

(4) Moisture Content

Moisture in the oil and insulation paper reaches an equilibrium state. The oil has different saturation levels of dissolved moisture at different temperatures. This saturation level increases with temperature, so moisture from the insulation paper enters the oil at high temperatures. When the oil temperature drops, some of the moisture in the oil will diffuse into the paper, causing a decrease in the water content in the oil.

(5) Flash point

The flash point is the lowest temperature at which the vapor of an oil mixture with air will ignite momentarily upon contact with a flame, under specified conditions of temperature, time, and flame size. The volatility of the oil product is actually related to the safety of transformer oil in the conditions of use and can be measured by the flash point.

The flash point and ignition point are not equivalent concepts. The flash point refers to the lowest temperature at which the oil gas produced is sufficient and when an external flame is added, the oil gas ignites momentarily; the ignition point is the lowest temperature at which an external flame can be maintained for 5 seconds when the oil gas continuously produced by heating. The sample is heated at a slow and constant rate under continuous stirring. At predetermined temperature intervals, a small flame is introduced into the cup while stirring is interrupted. The minimum temperature at which the flame causes the vapor on the specimen to flash fire is called the closed flash point.

(6) Water-soluble acid

Water-soluble acid refers to water-soluble mineral acids, primarily sulfuric acid and its derivatives, that are formed during the processing and storage of oil products. Transformer oil generally produces low molecular weight organic acids, such as formic acid and acetic acid, during the primary oxidation stage. These acids have good water solubility. When the water-soluble acid content in the oil increases (i. e., the pH value decreases) and there is water present in the oil, it can corrode solid insulation materials and metals, reduce the insulation performance of electrical equipment, and shorten the service life of the equipment.

Shake the test oil and an equal volume of distilled water vigorously to mix them. Extract the water phase from the mixture and add an indicator. Observe any color changes to determine the presence of water-soluble acids and water-soluble alkalis in the test oil. The results are represented by the pH value.

IV. Judgment standard of oil quality (see Table 3-23)

Table 3-23 GB/T 7595–2017 Quality Standard for In-service Transformer Oil

ID	Item	Equipment voltage class/kV	Quality Indicators		Test Method
			Oil before being put into operation	In-service oil	
1	Appearance		Transparent, free from impurities or suspended matter		Visual inspection for appearance
2	Water-soluble acids (pH value)		>5.4	≥4.2	GB/T 7598
3	Acid value/(mg KOH/g)		≤0.03	≤0.1	GB/T 264

Continued

ID	Item	Equipment voltage class/kV	Quality Indicators		Test Method
			Oil before being put into operation	In-service oil	
4	Flash point (closed)/°C		≥140 (#10, #25), ≥13.5 (#45)	Not less than 10 compared to the original measured value of new oil	GB/T 261
5	Moisture/(μL/L)	330–500 220 ≤110	≤10 ≤15 ≤20	≤15 ≤25 ≤35	GB/T 7600
6	Interfacial tension (25 °C)/(mN/m)		≥35	≥19	GB/T 6541
7	Dielectric loss factor/°C	500 ≤330	≤0.007 ≤0.010	≤0.020 ≤0.040	GB/T 5654
8	Breakdown voltage/kV	500 330 66–220 ≤35	≥60 ≥50 ≥40 ≥35	≥50 ≥45 ≥35 ≥30	GB/T 507 or DL/T 429.9
9	Volume resistivity (90 °C)/Ω·m	500 ≤330	≥6×10^{10}	≥1×10^{10} ≥5×10^{9}	GB/T 5654 or DL/T 421
10	Gas content in oil/% (by volume)	330–500	≤1	≤3	DL/T 423 or DL/T 450
11	Oil sludge and sediment/% (by mass)		<0.02 (negligible below)		
12	Analysis of Dissolved Gas Components in Oil by Chromatography		According to chapters 6, 7 and 9 of DL/T 596-2021		GB/T 17623

(1) The sampling oil temperature should be between 40 to 60 °C.
(2) DL/T 429.9 method uses a flat electrode, while GB/T 507 uses two types of electrodes: spherical and hemispherical. The different types of electrodes will yield different breakdown voltage values, which will impact the test results. The quality index is determined based on the measurements using the flat electrode

Module 3 Workshop

Ⅰ. Insulating Oil Dielectric Strength Test Steps and Precautions

When conducting a dielectric strength test on insulating oil using an insulating oil withstand voltage tester, the following procedure should be followed:

(1) Ground the instrument reliably.

(2) When the power is off, place the magnetometer in the oil cup.

(3) The "sample oil" must be placed in the laboratory for a period of time without damaging the original storage seal, until the oil temperature is close to room temperature before uncovering for the test. Before uncovering, gently shake the test oil to evenly distribute impurities, but do not produce bubbles. Before the test, wash the oil cup 2–3 times with the test oil.

(4) When the sample oil is loaded into the oil cup in a power-off state, it should flow down

slowly along the inner wall of the oil cup to reduce bubbles. During operation, it is not allowed to touch the electrodes, the inside of the oil cup, and the test oil with hands. After the oil cup is filled with test oil, it must be left still for 10–15 minutes before starting the breakdown test.

(5) In a power-off state, cover the electrode cover and close the high-voltage chamber lid.

(6) Turn on the power switch, and after the instrument displays the welcome screen, it automatically enters the main interface.

(7) Using the mouse, select operations such as breakdown test, withstand voltage test, view historical data, time setting, and PC communication.

Ⅱ. Dielectric strength test method for insulating oil

(Ⅰ) Operation method for breakdown test

(1) Set the test parameters, including initial settling time, number of tests, settling time, stirring time, and oil cup selection, as shown in Figure 3-42. The range for the initial settling time is 0 seconds to 9 minutes and 59 seconds, the range for settling time is 0 seconds to 9 minutes and 59 seconds, and the range for stirring time is 0 seconds to 99 seconds. The status of the three oil cups can be either "selected" or "not selected".

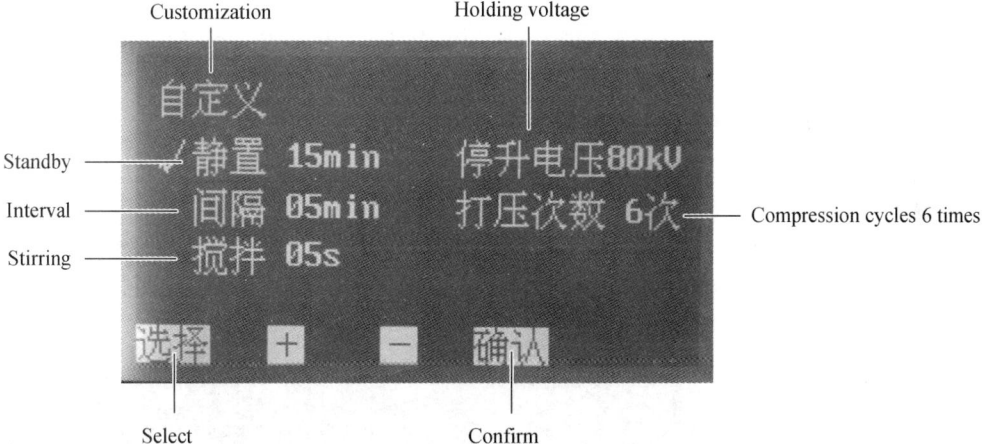

Figure 3-42　Test instrument parameter setting interface

① Settling time: default value is 15 minutes, range is 1 to 15 minutes (increment Δ = 1minute).

② Interval time: default value is 5 minutes, range is 1 to 10 minutes (increment Δ = 1 minute).

③ Stirring time: default value is 10 seconds, range is 5 to 90 seconds (increment Δ = 5 seconds).

④ Stop voltage: default value is 80 kV, range is 10 to 80 kV (increment Δ = 10 kV). When the instrument reaches the "stop voltage", it will stop increasing voltage and enter the hold state. If no breakdown occurs for 50 seconds, the instrument will default the current stop voltage as the dielectric breakdown voltage of the insulating oil.

⑤ Number of pressure cycles: default value is 6 times, optional range is 1 to 6 times (increment $\Delta = 1$ time).

After setting, press the "confirm" button to return to the start page, and press the "start" button to begin testing.

(2) Select "start test" and click "run" to start the test. The instrument will cycle through the sequence of increasing voltage until breakdown, stirring, settling, and then increasing voltage until breakdown again until the set number of tests is reached and the test stops.

(3) After the breakdown test is completed, the display screen of the instrument shows the test results, including breakdown voltage and average breakdown voltage, as shown in Figure 3-43.

(4) The operator can also save and print the test results as needed.

(Ⅱ) **The operating method of the withstand voltage test is as follows**

(1) Set the test parameters, including voltage, withstand time, and oil cup selection.

(2) Select "start test" and click "run". The instrument will increase voltage until it reaches the set withstand voltage. If a breakdown occurs during the voltage increase process, the test will end directly. If no breakdown occurs during the voltage increase process and the instrument reaches the set withstand voltage, it will stay at that voltage for the duration of the "withstand time".

(3) After the withstand voltage test is completed, the displayed test results include the withstand voltage, withstand time, and test result (OK for passing or NO for failing).

(4) The operator can also save and print the test results as needed.

The displayed content on the screen includes the total number of saved test groups, the serial number of the currently selected group, and the storage time. The operation for viewing historical data is as follows: Move the cursor to the "confirm" button to enter the test result viewing screen. The test result screen is shown in Figure 3-43. In these screens, choose to print, save or delete the data for that group.

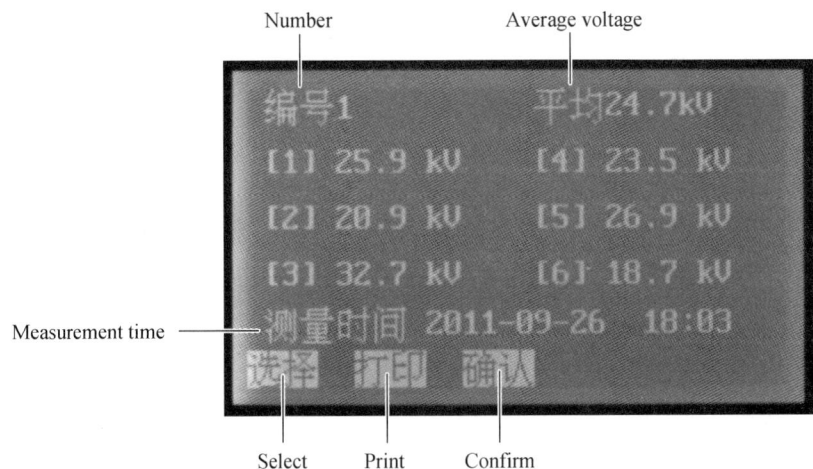

Figure 3-43　Withstand voltage test result

III. Insulation oil dielectric strength test precautions

Dielectric strength testing is quite sensitive to slight contamination of the sample, and it is easy for the sample to absorb moisture during sampling. Therefore, clean and dry sampling tools should be used to strictly follow the sampling method for electrical oil sampling. The test oil must be placed in the laboratory for a period of time without damaging the original storage and sealing before opening the cover for testing when the oil temperature is close to room temperature (the arbitration test should be performed between 15–20 °C). Before opening the cover, gently shake the test oil to mix impurities evenly, but do not produce bubbles.

1. Oil Cup Cleaning Method

(1) Use a clean cloth to repeatedly wipe the surface of the electrode and the electrode rod.

(2) Adjust the electrode gap with a standard gauge.

(3) Clean with anhydrous ethanol for 3–4 times, then dry with a hairdryer. Finally, clean with a test oil sample for 2–3 times.

2. Cleaning

Clean, and uncontaminated pair of tweezers or a magnetic glass rod to remove the stirring slurry. Clean the oil cup with the test oil for 2–3 times. During cleaning, pour the test oil from the electrode plate, then gently shake it back and forth to rinse the cup walls of the oil cup. When rinsing the oil cup, use the tweezers to hold the stirring slurry and rinse it in the rinsing solution.

3. Grounding

Before performing the voltage boosting operation, carefully inspect the connection of the circuit (including all connections and plugs) and the grounding of the ground wire.

4. Temperature and humidity

The test should be conducted under conditions where the humidity is not higher than 75%. The testing can only be performed when the temperature of the cooling oil is similar to room temperature.

5. Filling oil

When injecting the test oil into the oil cup, it should flow down slowly along the inner wall of the cup to avoid the formation of air bubbles. During operation, it is not allowed to touch the electrode, the inside of the oil cup, and the test oil with hands to prevent contamination of the test oil. After filling the oil cup with test oil, let it stand for 10–15 minutes with the glass cover on the sleeve.

6. Voltage test

The applied voltage must be increased from zero, with a rising rate of about 2 seconds per step, until the oil gap value is reached. Repeat this process 6 times, and the average value is taken as the measured value (the first time is not counted).

7. Current limitation during breakdown

In order to reduce the generation of carbon particles after the oil breakdown and to minimize the influence of impurities during breakdown, the impurities are stirred sufficiently in the oil between the electrodes after multiple breakdowns by water, and then allowed to settle for 5 minutes before repeating the test.

Module 4　Training

Cleaning method of oil cup during dielectric strength test of insulating oil:

_____
_____
_____

Worksheet 7 Switch Characteristic Test

Module 1 Operating Worksheet: Switch Characteristic Test

(Ⅰ) Test Name and Instrument	(Ⅱ) Test Objects
Switch characteristic test **Switch Characteristic Test Instrument**	Used for mechanical and electrical characteristic testing of vacuum switch, SF_6 switch, oil switch, disconnect switch, and GIS-combined electrical appliance at various voltage levels
(Ⅲ) Test Purpose	(Ⅳ) Measurement Steps
High-voltage switch mechanical characteristic testing includes parameters such as opening-closing time, three-phase synchronism, bounce count, and stroke. The purpose is to assess the reliability of the high-voltage switch and determine whether it can continue to be used, repaired, or replaced based on the test results	(1) Record the ambient temperature and humidity. (2) Connect the testing instrument to the GIS circuit breaker and install the sensors according to the requirements. (3) Turn on the power and set the relevant parameters of the instrument. (4) Adjust the output voltage of the power supply of the high-voltage switch mechanical characteristic testing instrument to the rated operating voltage as required, and perform opening and closing operations on the vacuum circuit breaker. (5) Record the opening and closing time, three-phase synchronism, speed, and closing bounce time for each phase
(Ⅴ) Precautions	(Ⅵ) Technical Standards
(1) The instrument must be reliably grounded. (2) The connection of the opening and closing control lines should be reliable. (3) Before testing, confirm that the switch is in the open position. (4) During testing, the operator should stand in a safe position and not touch the high-voltage switch. (5) After testing, the power should be disconnected in a timely manner to avoid other work while the equipment is still energized	(1) The synchronism of the circuit breaker's opening and closing operations should meet the following requirements: The difference in non-synchronization during closing between phases should not exceed 5 ms; The difference in non-synchronization during opening between phases should not exceed 3 ms; The difference in non-synchronization during closing between each contact within the same phase should not exceed 3 ms; The difference in non-synchronization during opening between each contact within the same phase should not exceed 2 ms. Unless otherwise specified by the manufacturer. (2) The parallel closing releaser should be able to operate reliably within the range of 85%–110% of its rated AC voltage or 80%–110% of its rated DC voltage; the parallel opening releaser should be able to operate reliably within the range of 65%–120% of its rated power supply voltage, and should not trip when the power supply voltage drops to 30% or lower of its rated value

Continued

(Ⅶ) Result Judgment	(Ⅷ) Digital Resources
(1) The closing and opening time and the synchronism between closing and opening should comply with the manufacturer's specifications. (2) The closing bounce time, unless otherwise specified by the manufacturer, should not exceed 2 ms	Circuit breaker opening and closing time test Circuit breaker opening and closing speed test Simultaneous measurement of circuit breakers Measurement of circuit breaker operating voltage

Module 2 Follow Me

Ⅰ. Meaning and Purpose of Circuit Breaker Mechanical Testing

Mechanical testing of the circuit breaker primarily consists of two parts: mechanical operation testing and mechanical characteristic testing.

Mechanical operation testing refers to various operational tests performed on the circuit breaker under no-load conditions (i. e. When there is no voltage or current in the main circuit). It is a test that verifies the mechanical performance and operational reliability of the circuit breaker.

Mechanical characteristic testing refers to the action time and motion speed of the circuit breaker contacts. It mainly includes parameters and test items such as the opening and closing time, opening and closing speed, synchronism of the main and auxiliary contacts during opening and closing, operating voltage of the opening and closing electromagnets, bounce, travel, and opening range.

Ⅱ. Mechanical Characteristic Testing of High Voltage Switches

The opening and closing speed, time, and degree of asynchronism of the circuit breaker, as well as the operating voltage of the opening and closing coils, directly affect the switching performance of the circuit breaker. Only by ensuring appropriate opening and closing speeds can the circuit breaker fully exert its ability to interrupt current, reduce contact wear caused by pre-breakdown during the closing process, and avoid contact burnout, oil spraying, and even explosions.

If the closing speed is reduced, when a short-circuit fault is closed, the hindering effect of the electromotive force that obstructs the contact closure will cause contact vibration or keep it in a stagnant state, especially when automatic reclosing fails, which may easily lead to explosions. On the contrary, if the closing speed is too high, the moving mechanism will be subjected to excessive mechanical stress, causing damage to individual components or shortening the service life. At the same time, due to strong mechanical impact and vibration, the contact bounce time will be extended. Serious asynchrony of the circuit breaker opening and closing will cause partial-phase connection or disconnection of the line or transformer, which may result in overvoltage that endangers insulation. Therefore, zinc oxide surge arresters are installed at the ends of transmission lines to limit overvoltage, as shown in Figure 3-44.

Figure 3-44　Installation of zinc oxide surge arresters at the ends of transmission lines to limit overvoltage

Some aspects of the mechanical characteristics of the circuit breaker are represented by the characteristic parameters of contact action time and motion speed. In mechanical characteristic testing, the most important parameters are generally the opening speed, closing speed, maximum opening speed, opening time, closing time, close-open time, open-close time, and synchronism of opening and closing. The contact motion speed during the circuit breaker opening and closing is an important characteristic parameter that affects the working performance of the circuit breaker, and the most important parameters are the opening speed and closing speed. Based on the opening and closing time of the circuit breaker and the stroke of the contact, the average speed of the contact motion can be calculated. The speed of the circuit breaker varies greatly during the entire motion process, so it is necessary to measure the actual contact motion speed of the circuit breaker.

III. Definition of Partial Time Parameters (As shown in Table 3-24)

Table 3-24 Definition of Time Parameters

ID	Time Parameter	Definition	Measurement Significance	Requirements
1	Opening time	The time interval from the moment when the opening command is received by the switch to the instant when the contacts of all poles are separated	Reduce the energy of the arc during the closing operation to prevent the welding of the contacts due to arcing	It should comply with the regulations of the manufacturer
2	Closing time	The time interval from the moment when the closing circuit is energized to the instant when the contacts of all poles are in contact for a switch that is in the open position	If the closing time is too short, the DC component may be too large when a system short circuit occurs, which may cause difficulties in closing the switch. If the closing time is too long, it will affect the stability of the system	It should comply with the regulations of the manufacturer
3	Open-close time	The time interval from the instant when the contacts of all poles are separated until the instant when the first pole makes contact during the automatic reclosing of the switch	The performance and capability of the switch in handling fault currents	It should comply with the regulations of the manufacturer. Generally, the closing time should be less than or equal to 100 ms and the opening time should be less than or equal to 80 ms. The small range can be controlled within: closing time of 35–70 ms and opening time of 20–60 ms
4	Close-open time	During the unsuccessful reclosing attempt or individual opening or closing operations of the switch, the time interval starts from the moment when the contacts of the first pole make contact until the instant when the contacts of all poles are separated during the subsequent opening operation	Check whether the switch is within the required range specified in the regulations. If it does not meet the requirements, it may easily cause overvoltage (operating overvoltage) during the operating process, affecting the quality of electric power and the stability of the power system	It should comply with the regulations of the manufacturer, which are generally not more than 60 ms for 126 kV and 252 kV, and not more than 50 ms for 363 kV and 550 kV
5	Simultaneity of opening and closing operations	The time difference between the three-phase disconnection and connection instants during the opening and closing operations of the switch, which is referred to as inter-phase synchronization, and the time difference between the disconnection and connection instants of each arc extinguishing unit of the same phase, which is referred to as intra-phase synchronization	The smaller the degree of non-simultaneousness, the better. Serious non-simultaneousness of circuit breaker opening and closing will result in non-full-phase connection or removal of the line or power-using equipment, which may produce over-voltage that jeopardizes the insulation of the equipment	Unless otherwise specified by the manufacturer, the circuit breaker's phase synchronization during closing and opening operations should meet the following requirements: the non-synchronization during closing between phases should not exceed 5 ms, the non-synchronization during opening between phases should not exceed 3 ms, the non-synchronization during closing between different poles of the same phase should not exceed 3 ms, and the non-synchronization during opening between different poles of the same phase should not exceed 2 ms

Continued

ID	Time Parameter	Definition	Measurement Significance	Requirements
6	Instantaneous speed of contact opening	The motion speed at which the moving contact and the stationary contact separate during the opening process of the switch	Reducing the speed of the opening process will increase the arcing time, causing damage or even welding of the contacts. When the internal pressure of the arc extinguishing chamber of the circuit breaker increases and a short circuit fault is cut off, it may cause explosion accidents	For different manufacturers and models, when there are no specific requirements, the instantaneous speed of the breaking point is calculated as the average speed within the first 10 ms after the opening operation
7	Instantaneous speed of contact closing	The motion speed at which the moving contact and the stationary contact make contact during the closing process of the switch	Reducing the speed of the closing process will cause the contacts to vibrate or stop moving due to the inhibiting effect of the electric force on the contact closure. If there is a short circuit fault during closing, it may cause an explosion	For different manufacturers and models, when there are no specific requirements, the instantaneous speed of the closing point is calculated as the average speed within the last 10 ms before the closing operation
8	Maximum opening speed	The maximum value of the average speed of a section during the opening process of the switch	If the maximum opening speed of the circuit breaker is too high, it will cause strong vibration, reduce the mechanical life of the circuit breaker, and cause misoperation of the relay protection. If it is too small, the speed of the opening process may not meet the requirements	calculated based on 10 ms

IV. The basis standard of the test item (As shown in Table 3-25)

Table 3-25 Application and standard of high voltage test testing instruments

Test Item	Production Standard	Test Standard Basis	Test Cycle
High-Voltage Switch Characteristic Test	DL/T 846.3–2017 General technical conditions for high-voltage test equipment. Part 3: Comprehensive test instrument for high-voltage switches	(1) According to GB 501502016 Electrical equipment installation Engineering electrical equipment handover test standard. (2) The test items for the oil circuit breaker should include the following: measurement of the opening and closing time of the oil circuit breaker, measurement of the opening and closing speed of the oil circuit breaker, measurement of the synchronism of the main contacts of the oil circuit breaker during opening and closing. (3) The test items for the vacuum circuit breaker should include the following: measurement of the opening and closing time of the circuit breaker's main contacts, measurement of the synchronism of the opening and closing, and measurement of the bounce time of the contacts during closing.	After the handover test and overhaul test of the circuit breaker, the relevant standards require the mechanical characteristic test

Continued

Test Item	Production Standard	Test Standard Basis	Test Cycle
High-Voltage Switch Characteristic Test	DL/T 846.3–2017 General technical conditions for high-voltage test equipment. Part 3: Comprehensive test instrument for high-voltage switches	(4) The test items for the SF_6 circuit breaker should include the following: measurement of the opening and closing time of the circuit breaker, measurement of the opening and closing speed of the circuit breaker, measurement of the synchronism and coordination time of the main and auxiliary contacts during opening and closing. (5) According to DL/T 596–2021 Preventive Test Regulations for Power Equipment: Test items, cycles, and requirements for the SF_6 circuit breaker and GIS: ① Measurements after major overhaul: speed characteristics of the circuit breaker, time parameters of the circuit breaker; test items, cycles, and requirements for the multiple oil circuit breaker and minimal oil circuit breaker. ② measurements after major overhaul: closing time and opening time of the circuit breaker, opening and closing speed of the circuit breaker, synchronism of the circuit breaker's contacts during opening and closing. ③ Test items, cycles, and requirements for the vacuum circuit breaker: measurements after major overhaul: closing time and opening time of the circuit breaker, synchronism of the opening and closing, bounce process during closing	After the handover test and overhaul test of the circuit breaker, the relevant standards require the mechanical characteristic test

Module 3　Workshop

Ⅰ. Mechanical Characteristics Test Wiring

The wiring is divided into instrument wiring and control line connection. When instrument wiring, please connect the instrument protective earth "⏚" to the site earth before performing other wiring and operations. After the test is completed, turn off the instrument power supply, remove other wires, and finally remove the grounding wire. The machine's disconnection line is a four-core sheathed wire with four colors: yellow, green, red, and black, corresponding to ABC and common ends respectively. The connection method is shown in Figure 3-45.

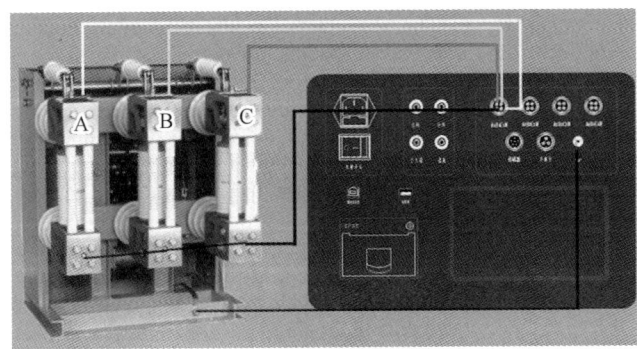

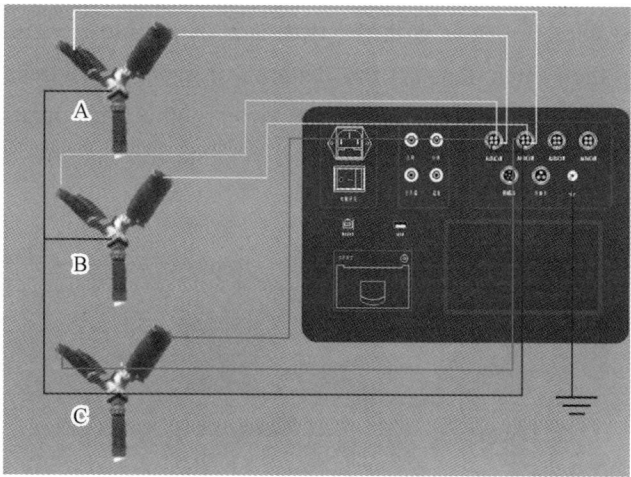

Figure 3-45 Wiring diagram of circuit breaker mechanical characteristics test instrument

When connecting the control lines, this machine provides a 3-core sheath wire as the internal power supply output wire (red for the closing wire, green for the opening wire, and black for the common power supply), and a 3-core sheath wire as the external action voltage acquisition wire. When the opening and closing control power supply is provided internally by the instrument, disconnect the control power supply in the control box of the tested switch (usually by removing the fuse that connects the control power supply in the control box to the control bus), but do not cut off the stored energy power supply of the switch mechanism. Then, follow the wiring diagram as figure 3-46. The internal power supply of the instrument can only provide DC current, and "internal triggering" should be used when using the internal power supply of the instrument. If the field switch has an AC operating mechanism, please use the "external triggering" mode.

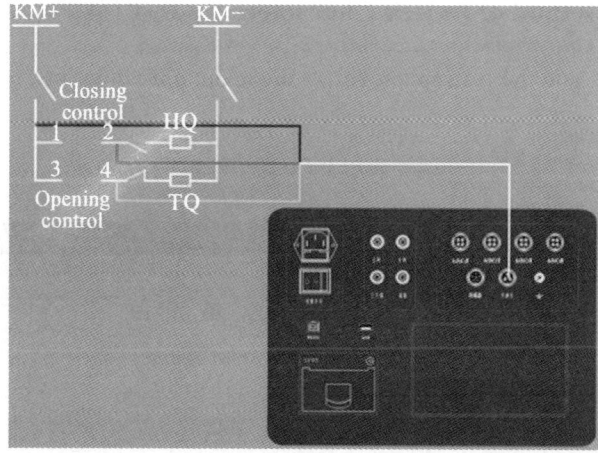

Figure 3-46 Control line wiring diagram of circuit breaker mechanical characteristic test

When using an external field power supply for opening and closing control, the "internal power supply" is not wired. When performing a single-closing test on the switch, connect the "external trigger" two wires together with the two ends of the closing coil; when performing a single-opening test on the switch, connect the "external trigger" two wires together with the two ends of the opening coil. When using an external power supply for operation, use the "external triggering" mode. The external triggering method is suitable for both AC and DC switches. The installation of the sensor is shown in Figure 3-47.

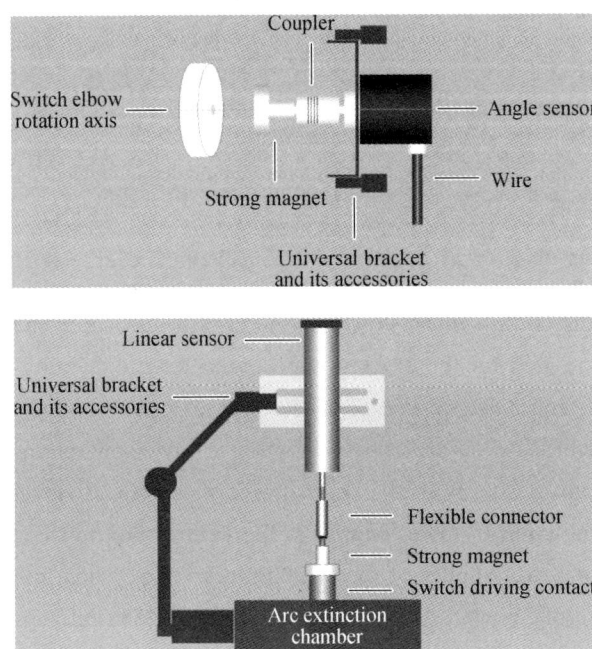

Figure 3-47 Sensor installation

Ⅱ. Operation Method of Switch Characteristic Test Instrument

Turn on the power, and the instrument enters the menu operation interface, as shown in Figure 3-48.

Figure 3-48 Human-Machine interface of switch characteristic test

(Ⅰ) Test Setting

Set the sensor type: There are several options such as linear resistance, rotary resistance, acceleration sensor, and photoelectric sensor. Set according to the sensor used. If there is no sensor, select "None".

1. Sensor Installation

Install a sensor three-phase linkage mechanism and choose three-phase linkage; install three sensors and choose three-phase simultaneous measurement.

2. Speed Testing

If speed testing is not required, this item can be turned off to shorten the test time and reduce the test intensity.

The speed definition: generally obtained by measuring the "time-travel characteristic curve" before closing and after opening for 10 ms, and then analyzing the corresponding speed value on the curve.

3. Travel Testing

When using a linear sensor or photoelectric sensor for testing, this item can be enabled to measure the travel distance. When using other sensors or not testing the travel distance, this item is set to close.

4. Labeled travel Setting

When measuring speed with a rotary or acceleration sensor, input the total travel value of the switch. When testing with a linear or photoelectric sensor, input the labeled travel value of the sensor.

5. Trigger mode

(1) Internal power supply trigger: Use the internal DC power supply of the instrument to operate the switch opening and closing.

(2) External power supply trigger: The internal DC power supply of the instrument does not work, and the switch action is operated by the on-site power supply (AC or DC). When the instrument makes a closing (opening) operation, the "external trigger" wiring of the instrument is directly connected in parallel to the closing (opening) coil. When the switch is operated, the instrument takes the voltage signal from the coil as the timing starting point.

(3) Auxiliary contact trigger: When there is no electrical signal on the coil, the test can be triggered by using the auxiliary contact method.

(4) Sensor trigger: Manual switch operation without electrical control mechanism, and no timing starting point. The measurement can be started by using the sensor action as the timing starting point.

(5) Manual switch: Measurement with manual switch. Just connect the break wire and perform the closing and opening tests. The instrument is in a waiting state. Then manually open and close the switch.

6. Test Duration

Refers to the duration of the operating voltage output by the internal power supply.

(1) 200 ms: Generally used for single opening and single closing tests of conventional switches, select a duration of 250 ms.

(2) 2,000 ms: Used when performing reclosing operations on general switches. It includes closing-opening, opening-closing, and opening-closing-opening.

(3) 20,000 ms: Generally used for contact switches with pre-action before the operation of opening or closing blades, which requires a long time for the operation of opening or closing blades.

7. Switch Types

(1) Metal Contact: Conventional switch with metal contacts, set to "Closed".

(2) Closing Resistor: For switches with closing resistors, if the closing resistance needs to be tested, set it to "Closing Resistor". If the closing resistance doesn't need to be tested, set it to "Closed".

(3) Graphite Contact: For Siemens graphite contact switches, it is necessary to set the switch to "Graphite Contact" during testing.

8. Pre-Closing/Closing Duration

Set this option for switches that require pre-closing or pre-opening. Generally, select "None".

Notes: After completing all the options, move the cursor to the "Confirm" button at the bottom of the screen and press the Confirm key to finalize all the settings.

(Ⅱ) Power Settings

Set the output voltage values for closing, opening, and reclosing according to the requirements of the switch. After setting, press "Confirm".

(Ⅲ) Status Detection

Check if the sensors are working properly and installed correctly. Also, verify if the switch is in the correct closed or open position. Ensure that the wiring is correct. Delete user-defined speed definitions.

(Ⅳ) Data Analysis

Figure 3-49 shows the test graphs and data for closing, opening, and closing-opening operations.

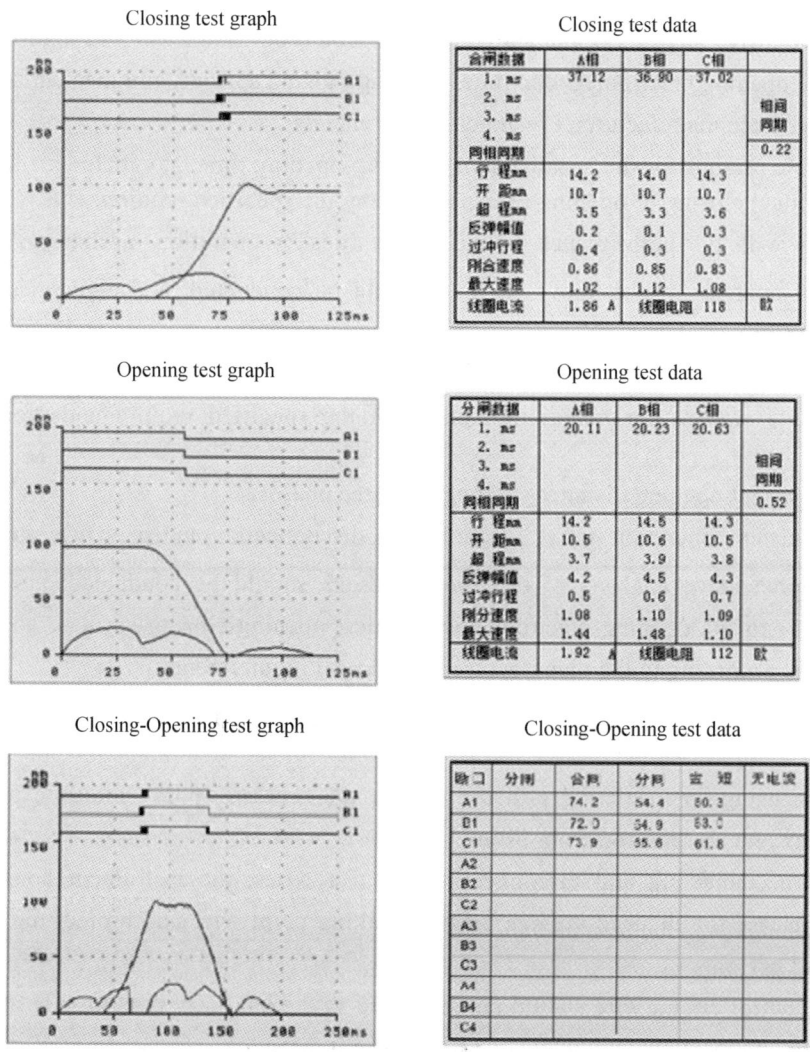

Figure 3-49 Test graphs and data for closing, opening and closing-opening operations

III. Analysis and Processing of Test Data for High-Voltage Switch Dynamic Characteristic Test Equipment

(1) Test results should be compared with the values given in the circuit breaker manual and meet the manufacturer's requirements.

(2) If there are any test data that do not meet the manufacturer's requirements in the test items, first check whether the wiring, parameter settings, instrument conditions, etc. Meet the test requirements.

(3) Possible reasons for the unsatisfactory opening time may include:
① Improper positioning of the opening electromagnetic pole rod and opening trip shaft.
② Fatigue of the opening spring.
③ Insufficient opening distance or overtravel.

A comprehensive analysis of the above reasons should be conducted, and adjustments should be made to the opening electromagnetic pole, opening-closing springs, and mechanism connecting rods according to the manufacturer's technical requirements.

(4) Possible reasons for the unsatisfactory closing-opening time may include:

① Individual closing or opening time not meeting the specified requirements.

② Issues with the performance of the circuit breaker operation mechanism's releaser. A comprehensive analysis of the above reasons should be conducted, and adjustments should be made to the individual closing or opening time or to the releaser according to the manufacturer's technical requirements.

(5) Possible reasons for non-compliance with the specified requirements for asynchrony include:

① Inconsistent opening distances among the three phases.

② Inconsistent operation times of the phase-disconnecting mechanism's electromagnetic poles. A comprehensive analysis of the above reasons should be conducted, and adjustments should be made to the opening electromagnetic poles, opening-closing springs, and mechanism connecting rods according to the manufacturer's technical requirements.

IV. Precautions for Mechanical Characteristic Testing

(1) When using an insulating pull rod to test the breaking point, multiple people should support it to prevent the pull rod from tilting.

(2) Before connecting and disconnecting the test wires, the instrument housing must be grounded to release the induced voltage on the breaking point wire and protect the safety of the instrument and personnel.

(3) The control circuit wire should be connected to the auxiliary contact side of the opening and closing coil, and it is strictly prohibited to connect it directly to the opening and closing coil to avoid burning out the coil.

(4) In actual operation, the circuit breaker is only activated when the control voltage is at the rated voltage. Therefore, the measured operating time of the circuit breaker is only accurate when the applied control voltage remains at the rated voltage. Generally, the higher the voltage applied to the opening circuit, the shorter the opening time of the circuit breaker, and vice versa. To prevent testing errors, it is necessary to ensure that the voltage applied to the opening and closing coil is the rated operating voltage.

(5) When installing sensors, the opening and closing operation pins of the circuit breaker should be inserted to prevent accidental activation of the circuit breaker during installation, which could cause mechanical injury to personnel. Whenever possible, install the sensor on the side of the moving lever closest to the movable contact to minimize the impact of gaps or non-linearity in the intermediate conversion section on testing accuracy. The sensor installation should be secure,

as any movement during the circuit breaker's operation can affect the accuracy of the test data (for rotary sensors, avoid the dead zone of the sensor).

(6) The mechanical characteristic test of the circuit breaker should be conducted under the rated SF_6 pressure and the rated operating mechanism pressure.

(7) When performing opening and closing operations on the circuit breaker, operators should communicate with each other and provide proper supervision. It is strictly prohibited to have personnel working near the mechanism box of the circuit breaker to prevent them from being mechanically injured.

(8) For the circuit breaker with different operating mechanisms, the adjustment of time and speed affects each other. Adjusting one parameter will change the values of other parameters. Therefore, after completing the adjustment, other parameters of the circuit breaker should be tested as well.

Module 4 Training

What specific content is tested in the characteristic test of high-voltage switches?

_____
_____
_____

Worksheet 8 Cable Fault Comprehensive Test

Module 1 Operating Worksheet: Cable Fault Comprehensive Test

(Ⅰ) Test Name and Instrument	(Ⅱ) Test Object
Cable Fault Location **Comprehensive Cable Fault Tester**	Various models and different voltage levels of power cable
(Ⅲ) Test Purpose	(Ⅳ) Measurement Steps
To locate high impedance flashover faults in cable, high and low resistance grounding faults, short circuits, cable breakage, poor contacts, and other faults	(1) Diagnose the nature of the cable fault and determine the type of cable fault. (2) Measure the distance to the cable fault point (rough measurement). (3) Trace the path of the cable fault. (4) Precisely locate the fault point (accurate measurement)
(Ⅴ) Precautions	(Ⅵ) Technical Standards
(1) Understand the basic information of the cable. (2) Ensure that the faulty cable is disconnected from power. (3) Discharge the cable before testing. (4) The high-voltage test of the cable should strictly follow the "Electricity Safety Work Regulations"	The middle joints and terminals of the cable are the most susceptible parts to faults in the cable. (1) The cable production process has problems, causing impurities, air gaps, etc. To enter the cable joint. When the cable is in operation, it will be subjected to a strong electric field, and impurities will become free, causing cable faults. (2) The metal of the cable joint is shielded and cannot effectively contact the ground, causing the grounding resistance of the cable joint to increase, breaking down the cable insulation and triggering faults
(Ⅶ) Result Judgment	(Ⅷ) Digital Resources
(1) If the insulation resistance between one or several cores of the tested cable and ground is low or less than 100 Ω, it can be judged as a low-resistance ground or short-circuit fault. (2) If the insulation resistance between one or several cores of the tested cable and ground is significantly different from the normal value and greater than 100 Ω, it can be judged as a high-resistance ground fault. This phenomenon is characterized by multiple instances of flashover, with a duration of a few seconds or minutes between each instance. (3) If the insulation resistance between the cores of the tested cable and ground is relatively high or within the normal range, it is necessary to stop the continuity test of conductors and check for any breakage. If a break is found, it can be judged as an open circuit fault. (4) Flashover faults are commonly observed during cable withstand voltage tests and typically occur at the cable terminals and middle joints. The insulation resistance of each phase can be measured using a megohmmeter during cable fault locating	**Cable Fault Comprehensive Test**

Module 2 Follow Me

Ⅰ. Overview of Cable Fault

The power cable is increasingly widely used due to its advantages of providing a safe and reliable power supply and contributing to urban beautification. The power cable is mostly buried underground, in harsh and complex environments, with multiple influencing factors. Once a fault occurs, it is very difficult to find out, often requiring several days or even longer for testing, which consumes a significant amount of manpower and resources and can cause incalculable power outage losses. Therefore, how to accurately and quickly locate cable faults has become an increasingly important issue.

Cable faults and the surrounding environment of cable installations are complex and variable. Testing personnel should be familiar with the cable routing and environment, accurately identify the nature of the fault, select appropriate instruments and measurement methods, and follow a certain working procedure in order to successfully locate the cable fault point.

Ⅱ. The basic structure diagram of power cable

The basic structure of a power cable consists of four parts: conductor, insulation layer, shielding layer, and protective layer. The conductor is the conductive part of the power cable, mainly used for transmitting electrical energy, and it is the main component of the power cable. The insulation layer electrically isolates the conductor from the ground and between different phase conductors, serving as a safety measure to protect against electric shock and ensure the transmission of electrical energy.

It is an essential component of the power cable structure. The shielding layer controls the electric or magnetic field within the cable, preventing interference with external equipment and preventing external electric or magnetic fields from entering the cable. The power cable with voltages of 15 kV and above generally has conductor shielding and insulation shielding. The protective layer is responsible for protecting the power cable from impurities and moisture intrusion, as well as preventing direct damage from external forces. The power cable is mainly classified into oil-immersed paper cable, rubber cable, and plastic cable based on the insulation materials. The plastic insulation materials include polyvinyl chloride (PVC) insulation, polyethylene (PE) insulation, cross-linked polyethylene (XLPE) insulation, polypropylene, etc. Among them, cross-linked polyethylene (XLPE) is the most widely used. Please refer to Figure 3-50 for details.

Ⅲ. Causes of Power Cable Failure

Various reasons can lead to different types of failures in the power cable, and being familiar with the causes of cable failures helps to quickly determine the location of the fault. There are mainly eight causes of failures, as shown in Figure 3-51.

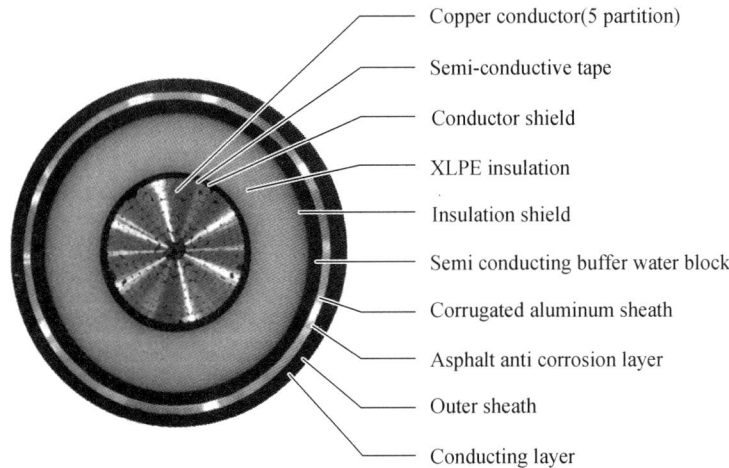

Figure 3-50　Structure diagram of cross-linked polyethylene cable

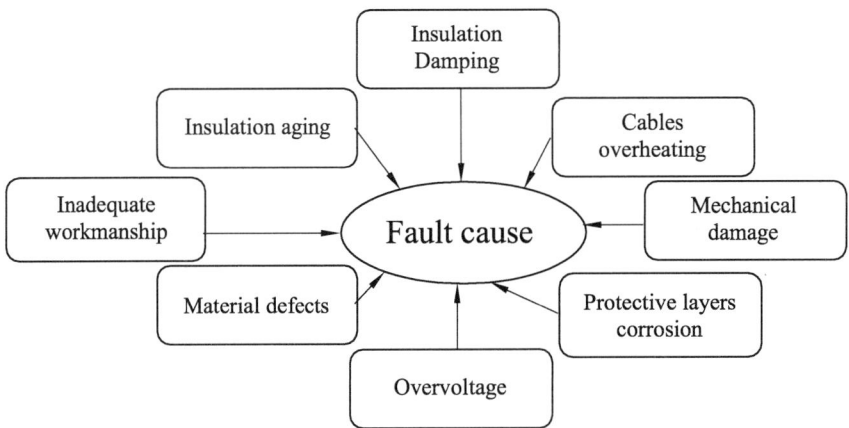

Figure 3-51　Eight common types of cable faults

(Ⅰ) Mechanical Damage

Mechanical damage is the main cause of cable faults and accounts for the largest proportion among various types of failures. During cable construction and maintenance, there are three main factors that lead to mechanical damage to the cable.

(1) External force damage is primarily caused by vibrations from construction activities such as pile driving and excavation near the cable path, or by heavy train operations. For example, the underground cable laid along the highway, railway, or subway may experience elastic fatigue and cracking of its outer sheath due to the severe vibrations or impact loads generated by passing trains.

(2) Installation damage factor is mainly caused by poor cable joint installation and laying processes, which can result in excessive stretching or bending of the cable under traction force, leading to damage to the insulation layer and shielding layer of the cable. Oversized cable stripping dimensions or excessively deep knife marks can also cause installation damage.

(3) Natural damage factor primarily occurs during the operation of the cable. It is caused by the excessive vertical force on the cable due to its own gravity or building settlement, resulting in deformation or even breaking of the cable. This can lead to fractures at cable joints, expansion of insulation gel inside the cable, and damage to the insulation layer and outer sheath of the cable, resulting in cable faults.

(II) Insulation Moisture

Insulation moisture is also prone to causing cable faults. Poor quality in the manufacturing process of intermediate or terminal cable joints can result in poor sealing and insulation damage with cracks, leading to moisture penetration into the insulation. Additionally, the metal sheath may develop holes or cracks due to corrosion or external punctures, which can easily lead to insulation moisture faults in the cable.

(III) Insulation Aging and Degradation

The insulation performance of the cable can undergo aging over time due to continuous electrical and thermal effects from both internal and external sources. This aging is characterized by a decrease in insulation strength and an increase in dielectric loss.

1. Chemical Change

Under the continuous action of the electric field, the gap within the insulation material of the cable undergoes chemical changes when ionized, resulting in the generation of corrosive chemical substances that affect the insulation performance of the cable. When the cable insulation is exposed to moisture, the presence of water can cause hydrolysis of internal fibers, leading to a decrease in insulation strength.

2. Material Fatigue

The insulation and protective layers of the cable gradually accumulate localized damage under the cyclic stresses from internal and external sources. This leads to a decrease in insulation performance, while the protective layer may develop small cracks and holes, resulting in cable faults.

3. Partial Overheating

Various losses during cable operation can cause the cable to heat up. When the cable is overloaded, the rate of temperature increase accelerates. Local overheating can result in accelerated insulation damage, leading to power outages or fires.

4. Over voltage

Atmospheric overvoltage and internal overvoltage in the power system can cause the electrical overvoltage on the cable insulation to exceed the allowable value, resulting in breakdown. Atmospheric overvoltage includes direct lightning overvoltage, induced overvoltage, and counterattack overvoltage, while internal overvoltage in the power system includes harmful overvoltage caused by improper power system operation or inappropriate coordination of grid parameters. Most outdoor cable terminal failures are caused by atmospheric overvoltage, and internal overvoltage can exacerbate existing cable defects and further enlarge them.

5. Poor Design and Manufacturing Processes

Quality issues such as waterproofing, mechanical strength, insulation materials, etc. It can lead to an unreasonable distribution of the internal electric field, insufficient mechanical strength and margins, and other factors that can affect cable quality. During the cable jointing process, poor sealing of lead, loose wire connections, excessive bending of core wires, moisture or air gaps in insulation materials, and other process quality problems can result in cable head failures.

6. Chemical Corrosion of Protective Layer

In areas where acid and alkali operations are present, the cable armor, lead sheath, or outer protective layer often suffer extensive corrosion damage in the soil due to chemical reactions with acids and alkalis.

7. Cable Overheating

There are multiple causes of cable overheating. Internally, cable insulation can experience local overheating due to the presence of internal gas voids, which leads to insulation carbonization. Externally, cable overheating can be caused by cable overload or poor heat dissipation. The cable installed in densely populated areas, cable trenches, and poorly ventilated cable tunnels, as well as the cable enclosed in dry conduits or located near thermal pipelines, are prone to overheating, resulting in accelerated insulation damage.

8. Material Defects

Material defects mainly manifest in three aspects:

(1) Issues in cable manufacturing, mainly including defects left by lead (aluminum) wrapping. During the insulation wrapping process, defects such as folding, cracking, openings, and overlapping gaps may occur on the paper insulation.

(2) Defects in the manufacturing of cable accessories, such as sand holes in cast iron parts, insufficient mechanical strength of porcelain components, non-compliance of other parts with specifications, or inadequate sealing during assembly.

(3) Poor maintenance and management of insulation materials, resulting in moisture, dirt, and aging of insulation materials in the production of cable intermediate joints and terminal heads.

Ⅳ. Judgment of the nature of the cable fault

Common cable faults include leakage grounding, short-circuiting (commonly known as "cable blasting"), and wire breakage. The main reasons for these faults are cable aging, external impact such as hitting, smashing, and squeezing, improper wiring processes, and protection failure, as shown in Table 3-26. The procedure for locating and dealing with cable faults is as follows: first, determine the nature of the fault, then locate the fault point, and finally, handle the situation according to the regulations.

Table 3-26 Types and Causes of Cable Faults

ID	Cable Fault Type		Fault Cause
1	Leakage Fault		The insulation resistance between core wires or between core wires and ground does not meet the requirements. Excessive leakage current between core wires or to the ground results in low insulation level of the cable, leading to leakage
2	Grounding Fault	Full Grounding ("Dead Grounding")	One phase core wire of the cable is grounded, and when measuring the insulation resistance between them using a megger (or multimeter), it shows zero resistance
3		Low Resistance Grounding	The insulation resistance value between one or several phase core wires of the cable and ground is below 500 kΩ
4		High Resistance Grounding	The insulation resistance value between one or several phase core wires of the cable and ground is above 500 kΩ, or even above 1 MΩ
5	Short Circuit Fault		There are completely short circuits, low-resistance or high-resistance short circuits; two-phase simultaneously grounded short circuits or two-phase direct short circuits; and three-phase short circuits or grounding
6	Open Circuit Fault		One or several phase core wires of the cable are disconnected, or a portion of a phase conductive core wire is broken
7	Flashover Failure		When the voltage of the cable reaches a certain value, flashover breakdown occurs between core wires or between core wires and ground. When the voltage decreases, the breakdown stops. In some cases, even if the voltage is increased again, no breakdown occurs, but it may happen after a certain period of time. This type of fault is characterized by having an automatically closed fault point
8	Cable Fire		After an interphase short circuit fault occurs, the protective devices such as fuses and overcurrent relays fail to function. The high temperature generated by the powerful short circuit current ignites the rubber skin of the rubber-coated cable, leading to a fire
9	Cable Damage and Cracking		This primarily occurs in the low-voltage rubber-insulated cable. Prolonged overload operation can lead to insulation aging and adhesion between the core wire insulation and the core wire, making interphase short-circuit accidents more likely to happen. The causes of these faults are not only related to improper selection of cable types and cross-sections, poor construction quality, and cable quality issues, but also often associated with cable management, operation, and maintenance

Ⅴ. Steps for Cable Fault Testing (see Figure 3-52)

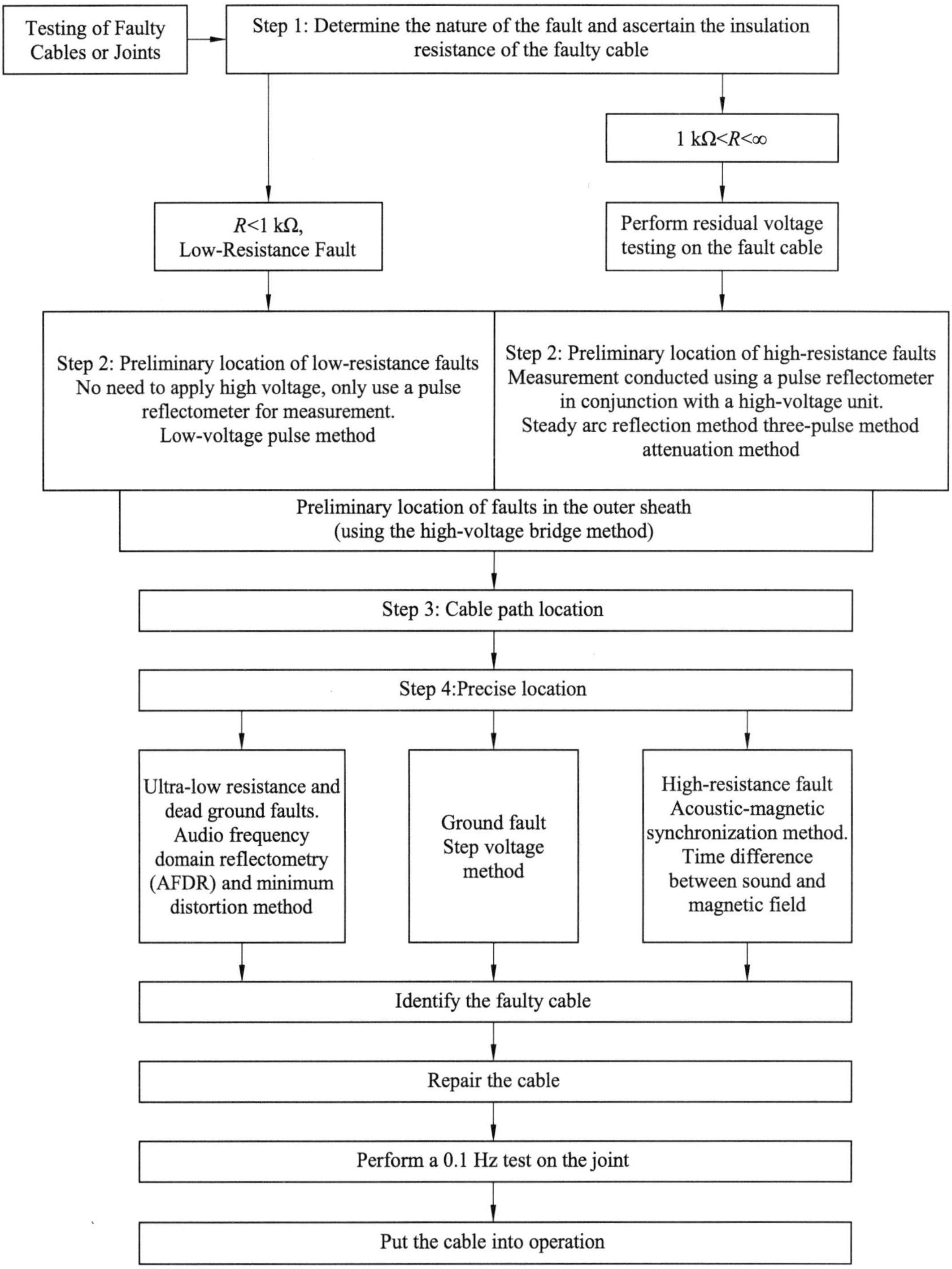
Testing of Faulty Cables or Joints
Step 1: Determine the nature of the fault and ascertain the insulation resistance of the faulty cable
1 kΩ<R<∞
R<1 kΩ, Low-Resistance Fault
Perform residual voltage testing on the fault cable
Step 2: Preliminary location of low-resistance faults No need to apply high voltage, only use a pulse reflectometer for measurement. Low-voltage pulse method
Step 2: Preliminary location of high-resistance faults Measurement conducted using a pulse reflectometer in conjunction with a high-voltage unit. Steady arc reflection method three-pulse method attenuation method
Preliminary location of faults in the outer sheath (using the high-voltage bridge method)
Step 3: Cable path location
Step 4:Precise location
Ultra-low resistance and dead ground faults. Audio frequency domain reflectometry (AFDR) and minimum distortion method
Ground fault Step voltage method
High-resistance fault Acoustic-magnetic synchronization method. Time difference between sound and magnetic field
Identify the faulty cable
Repair the cable
Perform a 0.1 Hz test on the joint
Put the cable into operation

Figure 3-52 Steps for cable fault testing

Ⅵ. Cable Fault Location

Wiring schematic of cable fault location is shown in Figure 3-53.

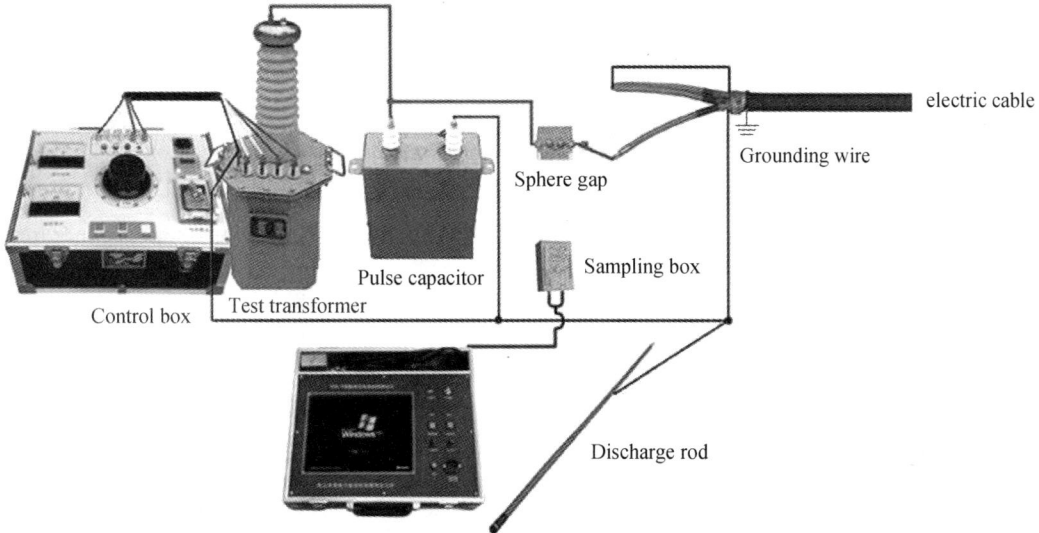

Figure 3-53 Wiring schematic of cable fault location

(Ⅰ) Acoustic Localization Method

1. Application range

The acoustic localization method is the primary point-based method for cable fault location. It is mainly used to measure high resistance and flashover faults, and can also be used for low resistance faults (except for metallic short circuits).

2. Locating the Fault Using Discharge Sound

By using the same high-voltage equipment as in flashover testing, the fault is triggered to discharge, resulting in mechanical vibrations during the fault gap discharge. These vibrations propagate to the ground, producing a "pa, pa" sound. This phenomenon can be used to accurately pinpoint cable faults.

For cable faults where the cable sheath has been burned through, the discharge sound at the fault point can be directly heard on the ground through headphones. However, for cable faults where the sheath is not burned through or when the cable is buried deeply, the discharge sound that can be heard on the ground is too faint. In such cases, a highly sensitive acoustic-electric transducer (pickup or piezoelectric chip) is required to convert the weak seismic waves on the ground into electrical signals. These signals are then amplified and processed to restore the sound through headphones.

Traditional cable fault locators rely on listening through headphones or observing the swing of a mechanical pointer to detect the audio signals generated by discharge at the fault points.

The sound waveform signals generated by discharge at the fault points are recorded. Analysis is performed to determine the intensity, frequency, attenuation, and duration of the signals in order to accurately identify the audio signals generated by discharge at the fault points.

The waveform of the sound signal generated by cable fault discharge, as received on the ground, depends on factors such as the size of the fault gap, whether the sheath is burned through, burial depth, and the surrounding medium of the cable. It can be challenging to perform accurate analysis. Generally, the waveform of the sound signal generated by cable fault discharge is a decaying cosine signal with a frequency between 200 and 400 Hz. The signal lasts for several milliseconds. Figure 3-54 provides a representative recorded waveform of the sound generated by cable fault discharge.

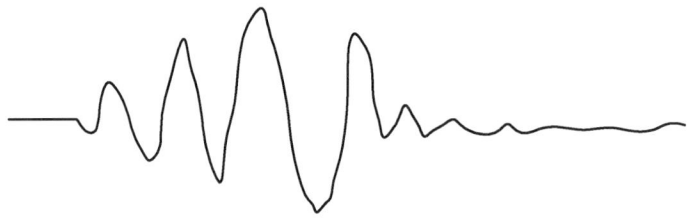

Figure 3-54　Waveform of discharge sound at the fault point

Intelligent fixed-point instruments have waveform memory and comparison functions. Multiple measurements can be taken at the same location to observe and compare the sound waveform signals recorded from multiple instances of fault discharge-triggered magnetic fields. If the sound signal is generated by fault discharge, there will be good consistency.

By utilizing this feature, it is possible to effectively eliminate the influence of environmental noise interference, especially when the sound signal generated by fault discharge is relatively weak. This method is highly effective.

3. Different types of fault connections

(1) Grounding fault: The high voltage impulse of a grounding fault is applied between the fault phase and the cable outer sheath. The vibrations generated by the fault gap discharge are transmitted through the outer sheath to the ground and can be easily detected.

(2) Phase-to-phase fault: In a phase-to-phase fault, the high voltage impulse is applied between two fault phases (with one phase connected to the outer sheath). The vibrations generated by the fault gap discharge are shielded by the cable insulation and protective layer, resulting in weaker seismic waves on the ground.

(3) Broken wire fault: The wiring diagram for an open-circuit fault without ground connection is shown in Figure 3-55. In this case, the fault phase needs to be grounded at the remote end to form a discharge circuit. Because the cable insulation and protective layer block the mechanical vibrations caused by the fault gap discharge at the breakpoint, the seismic waves received on the ground are weaker.

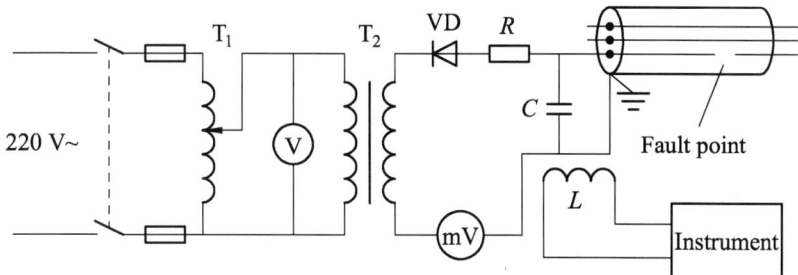

Figure 3-55 Wiring diagram for fixed point testing of open-circuit faults

4. Precautions

(1) The discharge energy at the fault point is related to the discharge current and the size of the grounding resistance. The resistance at the fault point should not be too low. Otherwise, the discharge energy will be small, and the locator may not detect the discharge sound. This is why the acoustic method is particularly suitable for high-resistance faults.

(2) It is advantageous to use large-capacity storage capacitors (2–9 µF) and increase the impulse voltage to increase the intensity of seismic waves generated by the discharge at the fault point. This facilitates the identification of the fault location.

(3) The time interval for sphere gap discharge is generally set to 2–10 seconds. If the discharge is too fast, the test equipment is prone to damage. If it is too slow, it is difficult to distinguish external interference. The discharge time is usually determined by adjusting the voltage of the voltage regulator and the size of the sphere gap. The discharge time interval of a dedicated high-voltage signal generator is controlled by a timing relay.

(4) During acoustic discharge testing, if the grounding is not good enough, there may be discharge phenomena between the cable insulation and the grounding part, leading to possible misjudgments. Therefore, special attention should be given to carefully identify the actual fault point, especially at the metal clamp of the exposed part of the cable.

In general, besides being able to hear sound at the fault point, there will also be vibrations. When touching the vibrating point, insulating gloves should be worn.

On the cable line between the power supply end and the fault point (including the bridging cable passing through iron pipes), during acoustic fixed-point testing, there may be induced voltages on the pipes and cable insulation, resulting in slight discharge sounds to the ground. These should be distinguished from the actual fault point. Generally, the sound at the actual fault point is louder and accompanied by vibrations.

(5) Fixed-point personnel and operators of high-voltage equipment can keep in touch through communication tools such as walkie-talkies, which make it easier to control the start-stop and time interval of gap discharge for high-voltage equipment. This is conducive to eliminating environmental noise interference and shortening the time required for fault localization.

(Ⅱ) Fixed-point method of synchronous reception of acoustic and magnetic signals

1. Synchronous reception of acoustic and magnetic signals enhances anti-interference capability

In practical testing, it is often difficult for people to identify the sound of discharge at the actual fault point due to environmental noise interference. The synchronous reception of acoustic and magnetic signals can improve identification capability.

When applying impulse high-voltage signals to the cable to cause discharge at the fault point, a circulating current is induced in the loop formed by the cable's outer sheath and the ground. This circulating current generates a pulse magnetic field around the cable. Since the pulse magnetic field generated by cable fault discharge is much stronger than general environmental electromagnetic interference, instruments can reliably detect the magnetic field signal. If a pulse magnetic field signal is received while listening to the sound signal, it can be judged that the sound is generated by discharge at the fault point, and the fault point is nearby. Otherwise, it can be considered as interference.

2. Identification of Pulse Magnetic Field Waveform

Modern intelligent fault localization instruments can record the pulse magnetic field signals generated by cable faults while recording sound signals. By identifying the characteristics of the pulse magnetic field, the influence of interference can be better eliminated. Comparing the initial polarity of the magnetic field waveform can determine the buried path of the cable while localizing the fault. The pulse magnetic field generated by cable fault discharge is generally a decaying cosine signal, and the period of the signal is related to factors such as the length of the cable and the surrounding medium. Its duration is approximately the same as the presence of high-voltage signals on the cable. Figure 3-56 shows a typical pulse magnetic field signal generated by a fault discharge point.

Figure 3-56　Pulse magnetic field waveform of fault point discharge

3. Fault Localization Utilizing Abnormal Changes in Pulse Magnetic Field

Since the magnetic field received on the ground is mainly generated by circulating currents between the cable's outer sheath and the ground, there is generally no significant change in the pulse magnetic field before and after the fault point. Therefore, the position of the fault point cannot be determined based on the waveform changes of the pulse magnetic field. However, in individual cases, such as severe breakdown of the cable's shielding layer at the fault point or when the fault point is located at a joint, there may be noticeable abnormal changes in the magnetic field on the ground caused by the fault discharge current. The position of the fault point can be determined based on this abnormal change in the magnetic field.

4. Fault Localization Utilizing Time Difference Between Magnetic and Sound Signals

During on-site testing, the sound of fault discharge is often heard, but the exact location of the fault point cannot be determined with certainty, especially when the cable is laid inside steel pipes or pipelines, which makes it even more difficult. This problem can be solved by detecting the time difference between magnetic and sound signals.

Since the magnetic field signal propagation speed is fast, the time required for the signal to propagate from the fault point to the instrument probe placement is usually in microseconds and can be ignored. On the other hand, the speed of sound propagation is slow, and the propagation time is in milliseconds. Therefore, based on the time difference between the magnetic and sound signals detected by the probe, the distance of the fault point can be determined, and the point with the minimum time difference can be measured, which is the fault point.

It should be noted that since it is difficult to know the speed of sound propagation in the surrounding medium of the cable accurately, it is not possible to determine the distance between the fault point and the probe accurately based on the time difference between the magnetic and sound signals.

Figure 3-57 shows the sound signal of fault discharge detected by the instrument probe at two points near the fault point. The instrument started recording the sound signal after being triggered by the pulse magnetic field signal. It is clearly evident from the figure that there is a time difference Δt_1 and Δt_2 between the occurrence of magnetic and sound signals. Since the second point is closer to the fault point, $\Delta t_1 < \Delta t_2$.

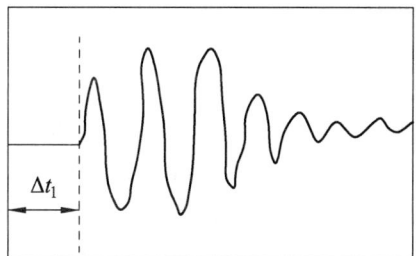
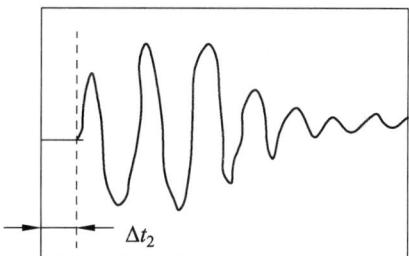

Figure 3-57 Recorded Sound Signal of Fault Discharge by Instrument

(Ⅲ) Audio Induction Method

1. Application Scope

The audio induction method is generally used to detect low-resistance faults with a fault resistance of less than 10 Ω. When the cable has a low grounding resistance, the sound of fault discharge is weak, and it is difficult to pinpoint the location using acoustic methods, especially for faults with metallic grounding where there is no sound of discharge at all. In such cases, the audio induction method is needed for special measurements.

The audio induction method can be used to test two-phase short circuit faults with grounding, as well as three-phase short circuit faults or three-phase short circuit faults with grounding, yielding satisfactory results. The absolute error in the position of the fault point obtained from the measurement is generally within 1–2 m.

For other types of faults, such as single or double-phase wire breakage, single-phase grounding faults, etc., if a special probe is used, they can also be accurately measured using the audio induction method.

2. Basic Principle of Fault Location

The basic principle of fault location using the audio induction method is the same as that for detecting underground cable paths using the audio induction method. During detection, an audio signal generator with a frequency of 1 kHz is used to pass an audio current through the cable under test, generating an electromagnetic wave. A probe is used on the ground to receive the electromagnetic field signal along the path of the cable being tested, and then send it to an amplifier for amplification. Then, the amplified signal is sent to headphones or indicator instruments, and the position of the fault point is determined based on the strength of the sound in the headphones or the magnitude indicated by the indicator instrument.

3. Methods for Fault Location

(1) Method for locating cable phase-to-phase short circuit (two-phase or three-phase short circuit) faults.

When using the audio induction method to locate the fault point of a phase-to-phase short circuit (two-phase or three-phase short circuit) fault, an audio current is passed through the short-circuited conductor core. A receiving coil is placed vertically or horizontally on the ground to receive the signal, which is then sent to a receiver for amplification. The magnetic field on the ground is mainly generated by the current flowing through the two conductors, and it varies with the torque of the cable. Therefore, when the probe moves along the path of the cable towards the fault point, a regular variation in sound can be heard. When the probe is positioned above the fault point, the sound is generally enhanced. As the probe continues to move forward from the fault point, the audio signal becomes noticeably weaker or even interrupted, as shown in Figure 3-58. Therefore, the point where the sound significantly weakens or interrupts is the fault point.

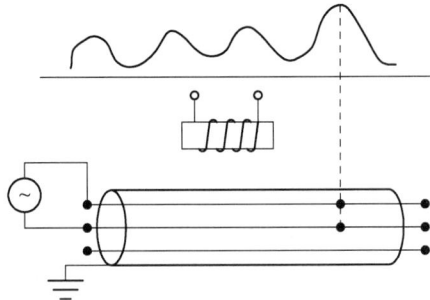

Figure 3-58　Detecting phase-to-phase short circuit faults in cable using audio induction method

The fault point location of phase-to-phase short circuit faults and phase-to-phase short circuit faults with ground connections is more sensitive when using the audio induction method.

(2) Method for Locating Single-Phase Ground Faults.

When locating the fault point of a single-phase ground fault, the audio signal generator is connected between the faulty phase conductor and the ground. As mentioned earlier, the magnetic field around the cable can be regarded as the sum of the magnetic field generated by the current I flowing between the conductor and the insulation, and the magnetic field generated by the current I' flowing between the metal sheath and the ground. The magnetic field generated by the current I' in the grounding circuit on the ground is much stronger than the magnetic field generated by the current I. Therefore, if a typical inductive coil is used to receive the signal, there will be no significant change in the signal sound heard at any point along the entire length of the cable, making it impossible to locate the fault point.

In this case, a special differential probe should be used. The differential probe has two coils, and the signals from the two coils are subtracted and sent to the instrument. When using the differential probe to detect along the cable path, the induced voltage generated by the current I' in the two coils is the same. Therefore, the signal sent to the instrument by the probe does not contain the magnetic field component generated by the current I'. It only reflects the magnetic field generated by the current I flowing between the conductor and the insulation.

Before the fault point, due to the twisted advancement of the conductor along the cable, the magnetic field along the cable varies. The differential probe can receive a weaker signal that changes along the cable before the fault point. After the fault point, there is no current I in the conductor, so the signal received by the differential probe is 0. The current in the cable conductor disappears at the fault point, resulting in a significant change in the magnetic field distribution near the fault point. The differential probe receives a stronger signal, which can be used to locate the fault point, as shown in Figure 3-59.

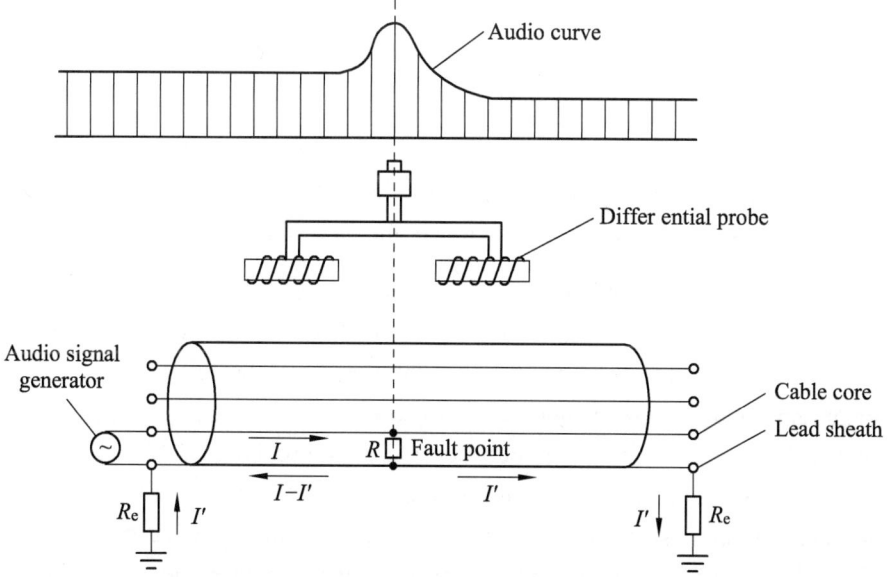

Figure 3-59　Principle of Audio Method for Locating Single-Phase Ground Faults

When using the differential probe, both magnetic rods of the probe should be parallel to the cable and detect signal along the path of the cable. They should not be offset or turned. If it is found that the differential effect is not working and there is significant stray interference, one of the two probes can be rotated 180 degrees in the horizontal plane.

In practice, it is quite difficult to use the audio induction method to measure ground faults, and often the fault point cannot be found. This should be noted.

(3) Several Issues to Consider When Locating Buried Cable Faults.

① When there is a ferromagnetic material around the cable, the signals received by the receiving coil may be strong, but this does not reflect the fault point.

② At cable joints, strong signals are often clearly received by the receiving coil.

③ If different parts of the cable are buried at varying depths, the strength of the signals received by the receiving coil will also vary. Shallowly buried sections will receive stronger signals.

Module 3 Workshop

Ⅰ. System Composition and Working Principle of Cable Fault Measuring Instruments

(Ⅰ) System composition

This system consists of four main parts: the cable fault measuring instrument host, locator, path finder, and high voltage signal generator.

1. Cable Fault Measuring Instrument (Distance Measuring Host)

The cable fault measuring instrument host is used to measure the nature of the fault, total length, and approximate location of the cable fault point.

2. Locator

The locator is used to determine the precise location of the fault point based on the approximate location determined by the host.

3. Path Finder

For the buried cable with unknown directions, a path finder is required to determine the direction. If the specific route of the cable is known, the path finder is not necessary.

4. High Voltage Signal Generator

It can be used as a high voltage source for cable fault testing at 35 kV and below, a DC withstand voltage test high voltage source, and with an external capacitor, it can also be used for energy storage impulse discharge testing.

Due to the differences in distances between the underground cable and the ground surface (the underground cable has a disk garden, a residual cable and a high, and low walking type), the distance measured by the host device is only the underground distance and may not be consistent with the distance on the ground surface. Therefore, we consider the cable fault measuring instrument host as a rough measurement device, and its test results provide only an approximate estimation. The precise location of the fault point can only be determined using the locator.

(Ⅱ) working principle

The host of the cable fault measuring instrument uses the Time Domain Reflectometry (TDR) principle. It emits an electrical pulse into the cable, which travels at a constant speed along the cable. When it encounters a change in cable impedance (fault point), the pulse reflects back. The host displays the emission and reflection changes of the electrical pulse in the time domain on the LCD screen, and the fault distance can be directly displayed on the screen.

Ⅱ. Steps for Cable Fault Location Finding

(Ⅰ) Instrument interface

Cable fault measuring instrument interface (distance measuring host) is shown in Figure 3-60.

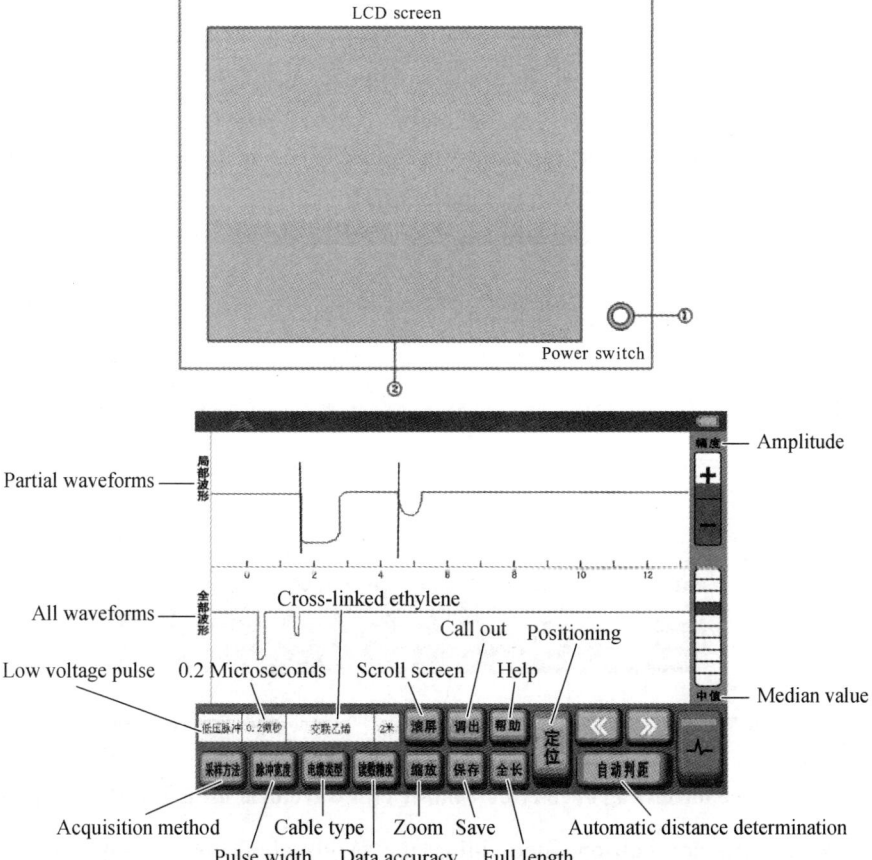

Figure 3-60 Schematic diagram of instrument panel and lcd screen menu display

(Ⅱ) Steps for Testing Low Resistance Grounding, Short Circuit, and Open Circuit Faults in Cable using Low Voltage Pulse Method

(1) No additional auxiliary equipment is required at this time. Simply connect a clip wire to

the input/output interface of the cable fault measuring instrument. Attach the red clip of the wire to the faulty core wire of the cable, and attach the black clip to the outer sheath ground wire of the cable, as shown in Figure 3-16.

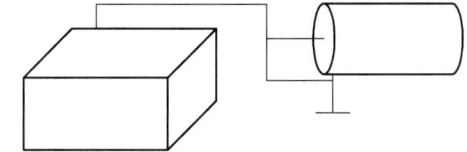

Figure 3-61　Low voltage pulse wiring diagram

(2) Turn on the power switch of the instrument, and a screensaver image will appear on the screen. Press it once to automatically enter the interface. At this time, the default state of the instrument is "Low Voltage Pulse Method". Depending on the type of cable being tested, its length, and the preliminary determination of the fault nature, choose the appropriate method. When set in the "Low Voltage Pulse Method", Perform velocity measurement and open historical files to review previous test results in this interface.

(3) After completing the device parameter settings, click the "Sampling" button, and the instrument will automatically generate a test pulse. This interface will display the full-length waveform of the cable (open circuit) (as shown in Figure 3-62) or the waveform of low resistance grounding (short circuit) fault (as shown in Figure 3-63).

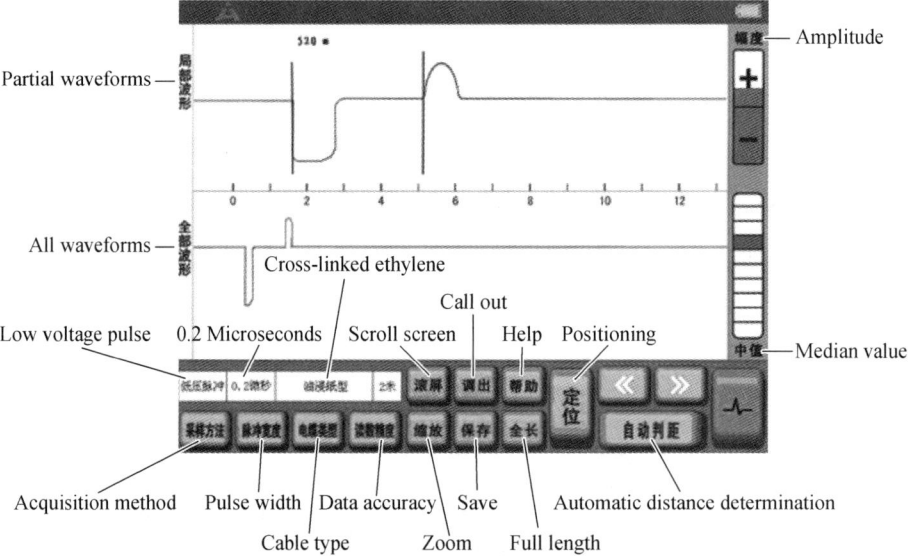

Figure 3-62　Interface for testing open circuit full-length waveform using low voltage pulse method

If the waveform is not well-operated, adjust the "Mid-value" and "Amplitude" settings, and observe the captured echo until the operator considers that the amplitude and position of the echo are suitable for analysis and positioning. The basic information of the instrument's parameter settings is also displayed at the bottom of the screen. Pay attention to the operation status below the screen during operation.

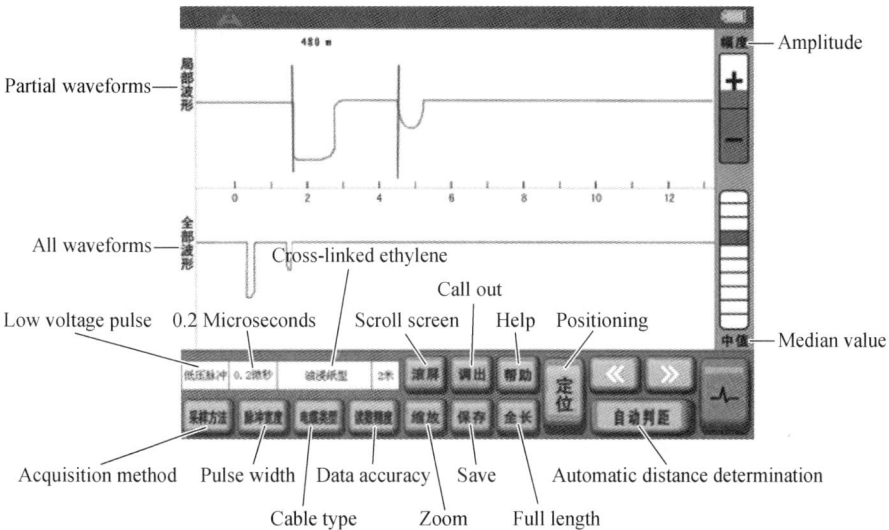

Figure 3-63　Interface for testing short circuit fault waveform using low voltage pulse method

(4) Waveform position reading distance.

It is relatively easy to determine the distance in low-voltage pulse testing. Just position the cursor at the starting points of the transmitted wave and the reflected wave, as shown in Figure 3-64.

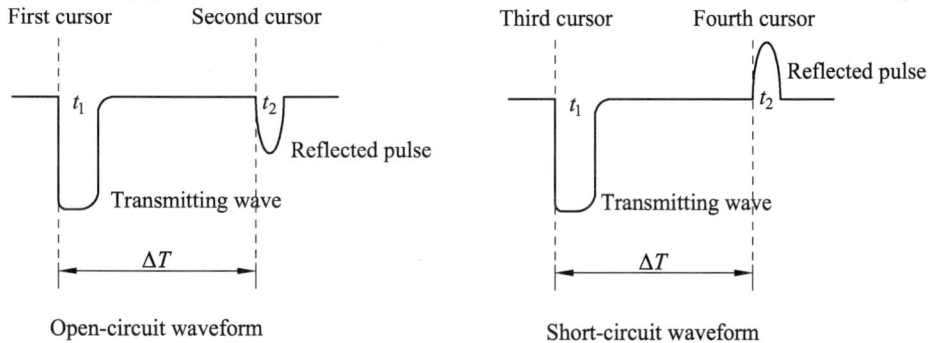

Figure 3-64　Low voltage pulse test waveform

(5) Save.

In many cases, it is necessary to preserve or keep the test results for future reference or comparison. In such cases, the "Save" function of the instrument can be used to save the waveform obtained in the instrument's database.

If the tester deems it necessary to save the test results, they can click the "Save" button and follow the prompts in the submenu for the operation.

(Ⅲ) **Testing High Resistance Leakage Faults (Including High Resistance Flashover Faults) in the cable with Impulse High Voltage Flashover Method**

The impulse high voltage flashover method is currently a popular traditional testing method for high resistance leakage faults in cable in China. The external circuit is relatively simple, but the difficulty lies in waveform analysis. Only after a large number of tests and some experience can one master it proficiently, but it is still an effective testing method.

After connecting the current sampler included with the instrument to the main unit with a signal cable, place it next to the ground wire between the cable and the high-voltage equipment. As long as the output voltage of the impulse high voltage generator is high enough, the fault point will be broken down under the impact of the impulse high voltage, and an electrical wave reflection will occur in the cable. The current sampler obtains the induced reflected wave of the current signal on the ground wire through magnetic coupling and transmits it to the LIXAAN-3000 cable fault measurement instrument for A/D sampling and data processing. The obtained waveform is displayed on the screen for fault distance analysis, as shown in Figure 3-65.

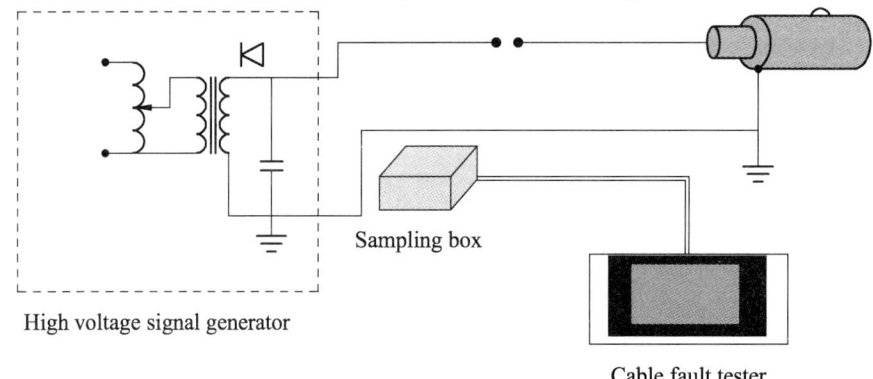

Figure 3-65   Wiring diagram of high voltage flashover testing method

After determining the cable type and sampling frequency, click the "Sampling" button to start the sampling process. Once the high voltage generator generates the impulse high voltage flashover, the instrument will automatically collect data and display the waveform, as shown in Figure 3-66.

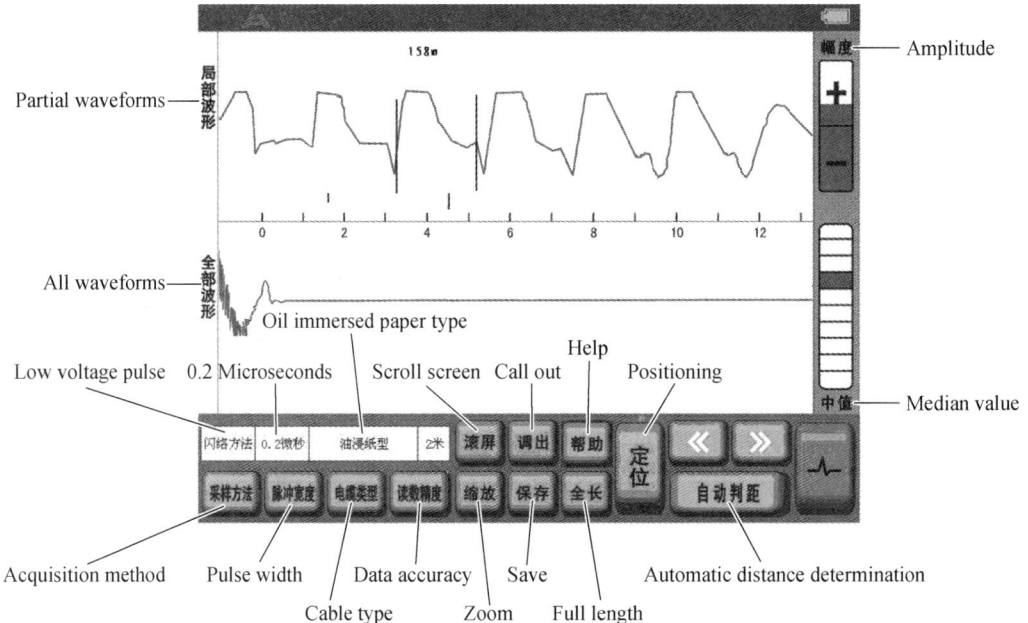

Figure 3-66   Waveform of high voltage flashover testing method

The red waveform at the top of the screen is the waveform after local amplification, and the blue waveform below is the complete testing waveform.

Once an ideal waveform is obtained, use the "Waveform Scaling" and cursor displacement or movement to calibrate the fault distance. The operation method is the same as the low-voltage pulse method.

(Ⅳ) Wave Velocity Measurement

To test the fault distance more accurately, it is often necessary to verify (test) the wave propagation speed of the cable.

(1) First, select a known length of the cable to be tested. If the length of the cable to be tested is not known, this cable also can be used for velocity measurement.

(2) After entering the setting interface of the instrument, press the "Sampling Method" button and select the "Velocity Measurement" option. Choose an appropriate sampling frequency and pulse width. Connect the measurement clip wires of the instrument to the core wire and the outer sheath of the cable to be tested. Press the "Total Length" button to prompt a dialog box where you can enter the cable length value, then press "OK". Click the "Sampling" button and the instrument screen will display the waveform of the open circuit test using low-voltage pulses. The instrument will automatically display the selected cable's test velocity based on cursor positioning, as shown in Figure 3-67.

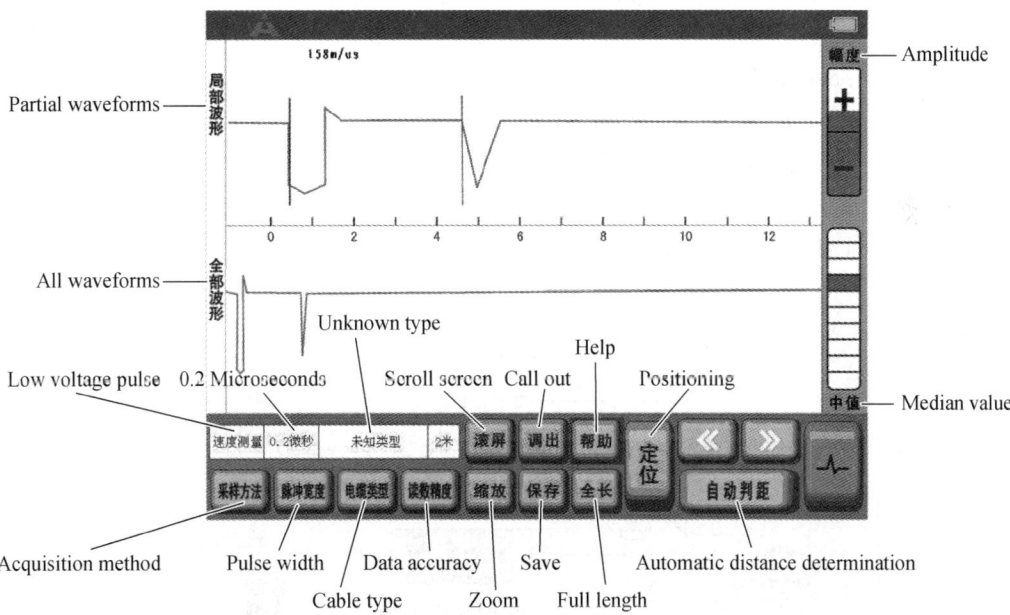

Figure 3-67 Interface diagram during speed measurement

Ⅲ. Cable Fault Testing Operation Steps

(Ⅰ) Panel Introduction

High voltage generator control instrument operation panel is shown in Figure 3-68.

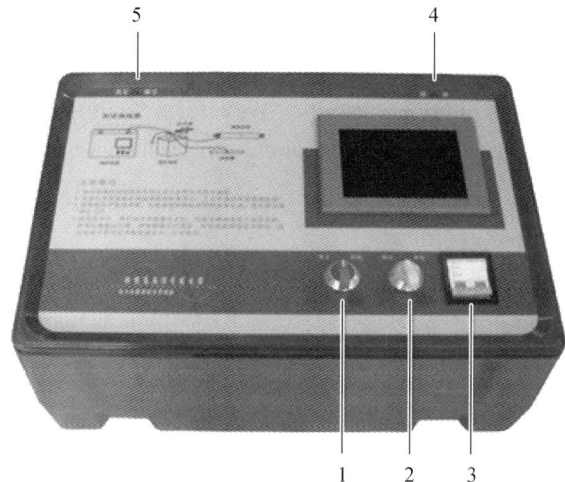

1–Power control switch; 2–Voltage up/down switch; 3–Power air switch;
4–Power socket and high voltage source grounding terminal; 5–High voltage output interface.

Figure 3-68　High voltage generator control instrument operation panel

(Ⅱ) Cable Fault Testing Method

1. Wiring for Flashover Method

(1) Connect the high voltage output terminal of the high voltage power supply to one terminal of the high voltage pulse capacitor. Then connect this terminal of the capacitor to one end of the discharge ball gap, and connect the other end of the gap to the faulty phase of the cable. The other phases and the outer armor should be grounded together. The other terminal of the high-voltage pulse capacitor is directly connected to the system ground.

(2) The grounding terminal of the high-voltage power supply is directly connected to the grounding terminal of the capacitor.

(3) After connecting the current sampling box to the sampling host, place it next to the capacitor's ground wire. Refer to Figure 3-69.

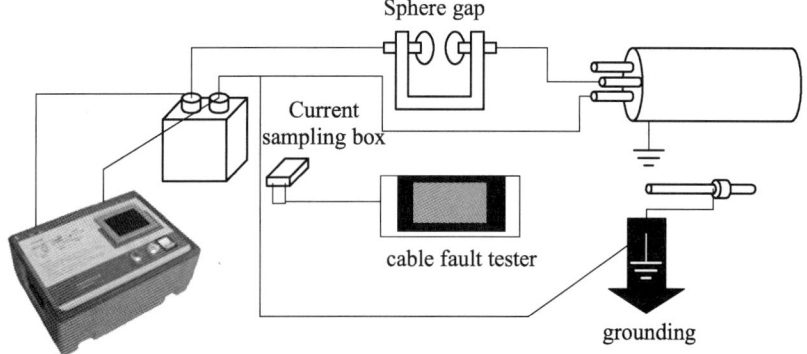

Figure 3-69　The wiring diagram of flashover current sampling

2. Wiring for Three-stage Multiple Pulse Testing Method

(1) Connect the high voltage output terminal of the high voltage power supply to one terminal

of the high voltage pulse capacitor. Then connect this terminal of the capacitor to the high-voltage input of the central control unit. The high voltage output of the central control unit is connected to the faulty phase of the cable. The other phases, outer armor, and the ground wire of the central control unit should be grounded together.

(2) The grounding terminal of the high-voltage power supply is directly connected to the grounding terminal of the capacitor.

(3) Connect the sampling host to the sampling port of the central control unit. Refer to Figure 3-70.

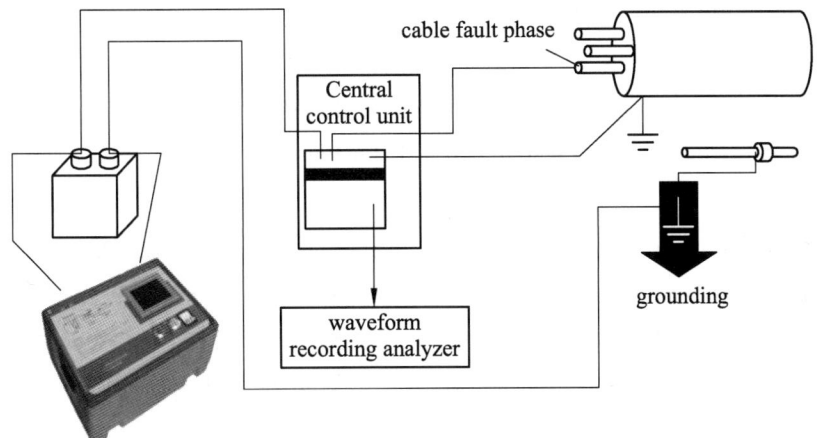

Figure 3-70　Wiring diagram for three-level pulse testing

3. The wiring for fault location

The wiring for fault location is the same as the wiring for the flashover method, except that sampling is not required. Refer to Figure 3-71.

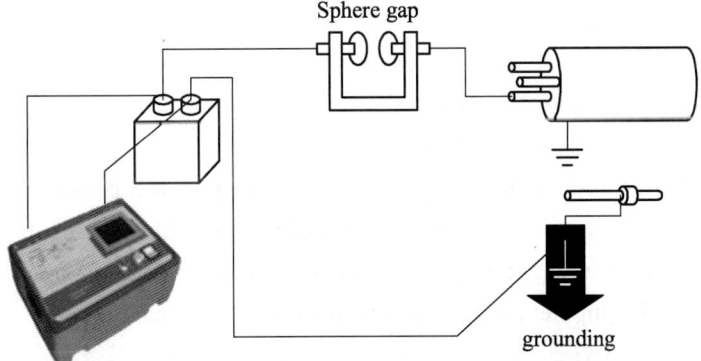

Figure 3-71　Wiring diagram for fault point location

(Ⅲ) Instrument Operation Steps

1. Wiring

(1) Connect the high-voltage output terminal of the high-voltage power supply to the

high-voltage capacitor.

(2) Connect the ground terminal of the high-voltage power supply directly to the ground terminal of the high-voltage capacitor, and then connect the capacitor's ground to the protective ground.

(3) Ensure that the control switch is in the "stop" position. Insert the power cord into the power socket of the high-voltage power supply, and then turn on the air switch The main control board will start working normally, and the LCD screen will display the current working status and the voltage status of the high-voltage capacitor.

2. Start

Rotate the start/stop control knob to the "start" position. At this time, the voltage preset value in the working status bar on the LCD screen will be highlighted. The power supply enters the standby state.

3. Test operation

Step up: Rotate the voltage adjustment knob continuously to the step-up position, and the value in the voltage preset bar will gradually increase. Release the voltage adjustment knob to its original position, and the power supply enters the boosting working state. It charges the external capacitor and boosts it to the preset value.

Step down: Rotate the voltage adjustment knob continuously to the step-down position, and the value in the voltage preset bar will gradually decrease. Release the voltage adjustment knob to its original position, and the power supply enters the reducing working state. If the voltage of the external capacitor is higher than the preset value, use a discharge rod to discharge the capacitor to the preset voltage.

4. Shutdown

(1) Rotate the start/stop control knob to the "stop" position. At this time, the voltage preset value in the working status bar on the LCD screen will be grayed out, and the power supply cannot perform stepping-up or stepping-down operations. The voltage meter on the LCD screen can still display the voltage value of the external capacitor.

(2) Use a discharge rod to discharge the electricity from the powered equipment or test sample while monitoring the voltage meter on the LCD screen. Continuous discharging until the voltage indication reaches zero. Then, use a grounding wire to directly touch the high-voltage terminal of the powered equipment or test sample and hang the discharge rod.

(3) Switch off the air switch of the high-voltage power supply.

(4) Remove the wiring.

Ⅳ. Cable Fault Location

(Ⅰ) Introduction to the Synchronized Receiver Locator Panel.

The Synchronized receiver locator panel is shown in Figure 3-72.

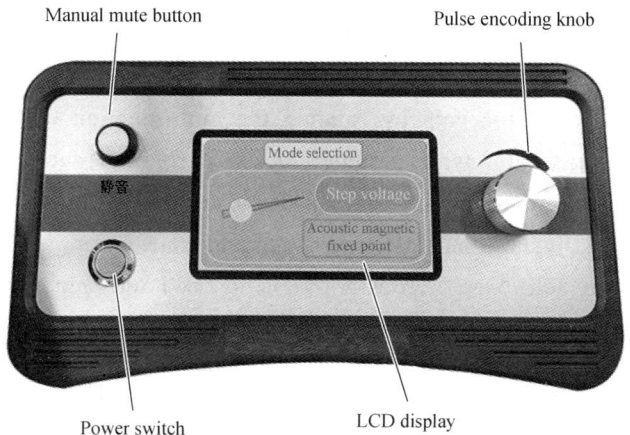

Figure 3-72 Synchronized receiver locator panel

(Ⅱ) Operation Method

(1) Turn on the power switch. After a brief startup animation, it is the mode selection interface. At this time, rotate the multifunction knob clockwise to point the pointer to "Acoustic-Magnetic Point", or rotate it counterclockwise to point the pointer to "Step Voltage". Once selected, press the multifunction knob to enter the corresponding working mode, as shown in Figure 3-73.

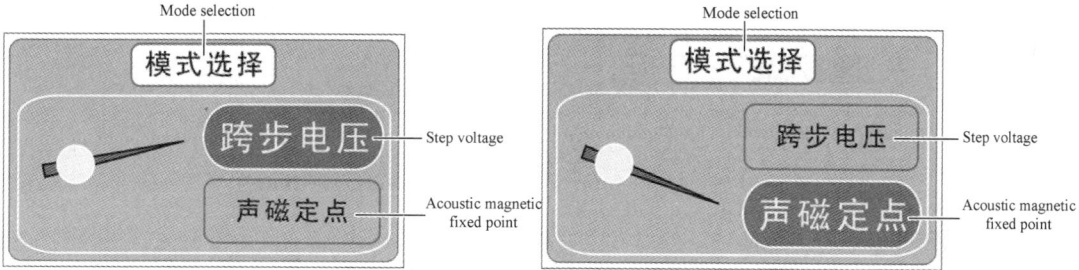

Figure 3-73 Two working modes of the synchronized receiver locator

(2) Select the "Step Voltage" mode, and the main interface of the "Step Voltage" mode will appear, as shown in Figure 3-74.

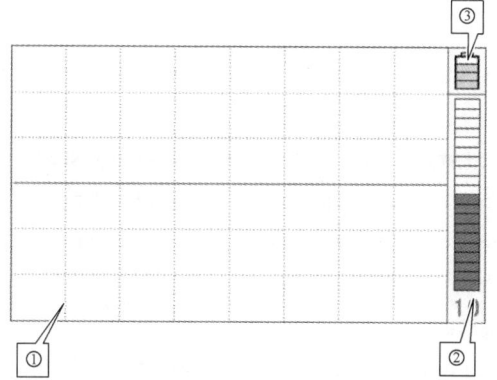

Figure 3-74 Main Interface of the Step Voltage Mode

① Waveform Display Area: The waveform display area continuously refreshes the collected waveform in real time, and users can judge the fault direction by changes in the signal starting edge and signal amplitude.

② Gain Display: Adjust the gain by rotating the multifunction knob. Clockwise rotation increases the amplification gain, while counterclockwise rotation decreases it. There are 20 subdivisions that can be adjusted, with internal automatic attenuation.

③ Battery Display: Displays the battery power in real-time.

(3) After booting up, select the "Acoustic-Magnetic Fix Point" mode to enter, and the main interface of the "Acoustic-Magnetic Fix Point" mode will appear, as shown in Figure 3-75.

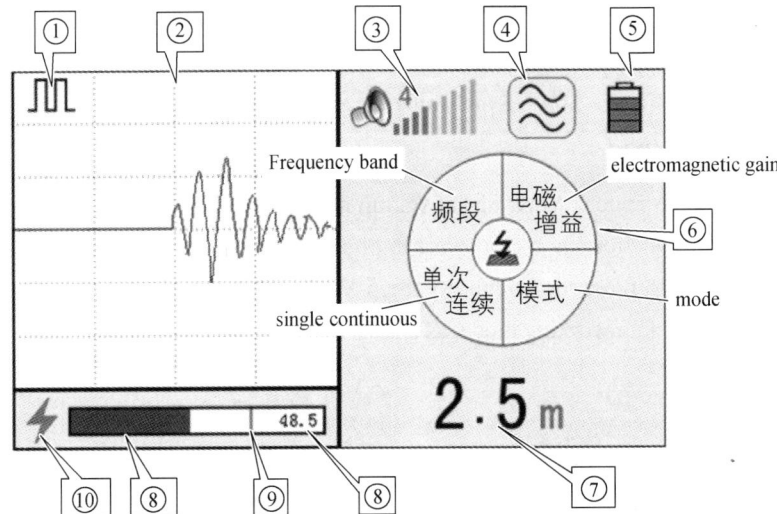

Figure 3-75　Main Interface of the Step Voltage Mode

① Mode Indicator

Displays the instrument's working mode, including single and continuous modes.

② Waveform Display Area

Displays the waveform of the last audio signal collected. When there is no electromagnetic interruption for a long time, the waveform automatically clears. In single mode, the background will automatically add a distance axis, and rotating the multifunction knob can perform page-turning operations to facilitate viewing of waveform information on a large scale.

③ Audio Gain

Under the main interface status, the audio gain can be adjusted by rotating the multifunction knob clockwise to increase or counterclockwise to decrease. After pressing the mute button on the panel, the instrument outputs silence. Pressing it again or changing the audio gain will restore normal audio output.

④ Frequency Band Display

Displays the current frequency band of the instrument, including four bands: full band, low band, medium band, and high band, as shown in Figure 3-76. The frequency band can be changed in the main menu frequency band column.

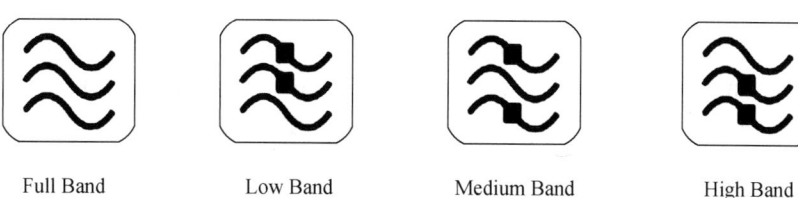

Figure 3-76 Frequency Band Display

⑤ Battery Display

Real-time display of the remaining battery power of the instrument.

⑥ Main Menu and Submenu Display

Displays the main menu, allowing selection and switching of the main menu by pressing and rotating the pulse encoder. It also enables the adjustment of internal parameters within the submenu.

The main menu disc display interface when the pulse coding knob is not pressed under the main interface, as shown in above Figure 3-77.

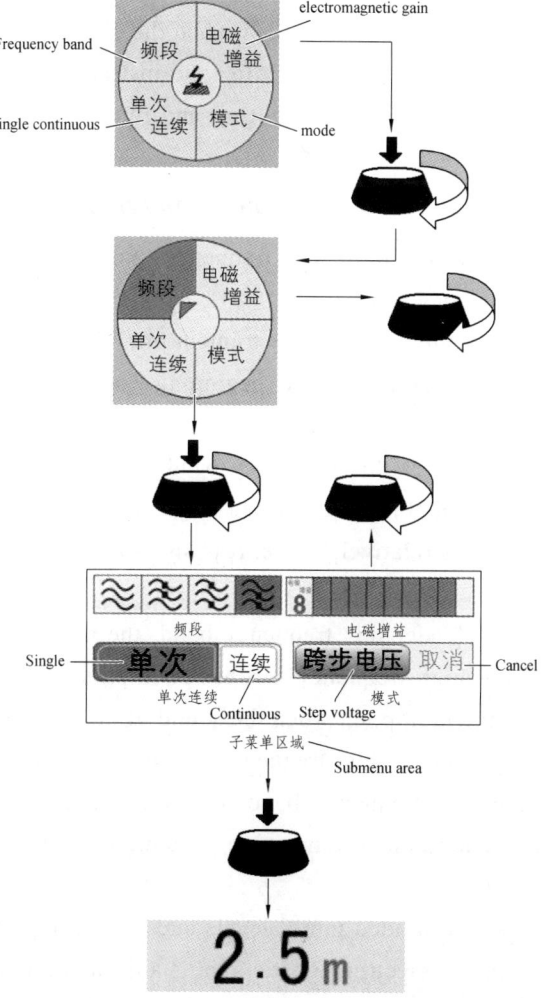

Figure 3-77 Main menu and submenu

The display interface when pressing the pulse coding knob as shown in the middle of Figure 3-77 (At this point, menus can be switched by rotating the pulse encoder knob).

After selecting a menu, press the pulse encoder key again to enter the menu. At this point, the corresponding submenu for the selected frequency band, electromagnetic gain, and single mode will appear in the distance display area. After calling up the submenu, adjust the contents of the submenu by rotating the pulse encoder knob, as shown in the below figure 3-77. Pressing the pulse encoder knob again confirms the exit. At this point, the submenu returns to the distance display.

⑦ Electromagnetic Signal Strength: After receiving an electromagnetic interruption, it is automatically filled based on the electromagnetic strength. At the same time, the current collected electromagnetic signal strength is displayed on the right side of the electromagnetic bar, ranging from 0 to 99.9. The display will automatically clear after approximately 3 seconds.

⑧ Maximum Electromagnetic Strength Indicator: Records the maximum electromagnetic strength under the current electromagnetic gain. It will be automatically reset to zero if no electromagnetic signal is received for a long time or when the electromagnetic gain is changed.

⑨ Electromagnetic Interruption Indicator: After the instrument receives an electromagnetic interruption, the indicator will automatically light up and remain lit for about 3 seconds before automatically clearing.

(Ⅲ) **Introduction to various modes of acoustic and magnetic fix points**

(1) Frequency Band: Divided into full frequency band, low-frequency band, medium frequency band, and high-frequency band.

① Full frequency band: At this time, the instrument provides the widest operating frequency band and is suitable for fixing points at the beginning stage with relatively small external interference.

② Low-frequency band: Under this frequency band, the noise in the high-frequency range will be greatly attenuated. It is suitable for use when the fault point is relatively far away or when the soil or sand above the cable is relatively loose. It is also suitable for use when the sound of the spark is relatively "muffled".

③ High-frequency band: Under this frequency band, the noise in the low-frequency range will be greatly attenuated. The high-frequency part has good passability and is suitable for use on relatively hard roads or near the fault point when the sound of the spark is louder.

④ Medium frequency band: This frequency band is a frequency band between the low-frequency band and the high-frequency band. Users can choose according to the different responses of the discharge sound in the low-frequency and high-frequency parts at this time.

(2) Electromagnetic gain.

The electromagnetic gain is divided into 9 levels and can be adjusted according to different on-site conditions. Every time the electromagnetic gain is adjusted, the maximum electromagnetic strength distance will be displayed.

Notes: During use, if the electromagnetic part keeps triggering, it is necessary to appropriately reduce the current electromagnetic gain.

(3) Mode switching.

During acoustic and magnetic fix points, the step voltage mode needs to be switched quickly, select the mode column in the menu, then select the step voltage, press confirm, and the working mode will switch to the step voltage mode quickly. In step voltage mode, if acoustic and magnetic fix points mode needs to be switched quickly, simply press the pulse code knob.

(4) Single/Continuous.

In the main menu, select Single/Continuous. When selecting the single mode and confirming to exit, the background of the waveform display area will show the distance axis. The instrument will only respond once to the spark mode and will not respond again. By rotating the multifunction knob, waveform information can be switched to observe, with a total observable range of 15 m (Notes: Audio gain cannot be adjusted at this time, and the multifunction knob becomes a waveform display page switch knob). If it needs to respond to electromagnetic interruptions again, enter the main menu, select Single/Continuous, select Single mode, and press confirm again to respond to electromagnetic interruptions again.

When selecting Continuous mode, the instrument will automatically respond to electromagnetic interruptions and refresh the display information each time. The waveform display in this mode is a compressed waveform of 15 m, so observing the general information of the waveform without much operation is allowed. At this time, adjust the audio gain by rotating the multifunction knob in the main interface state. If it needs to observe detailed waveform information, enter the Single mode to observe as shown in Figure 3-78.

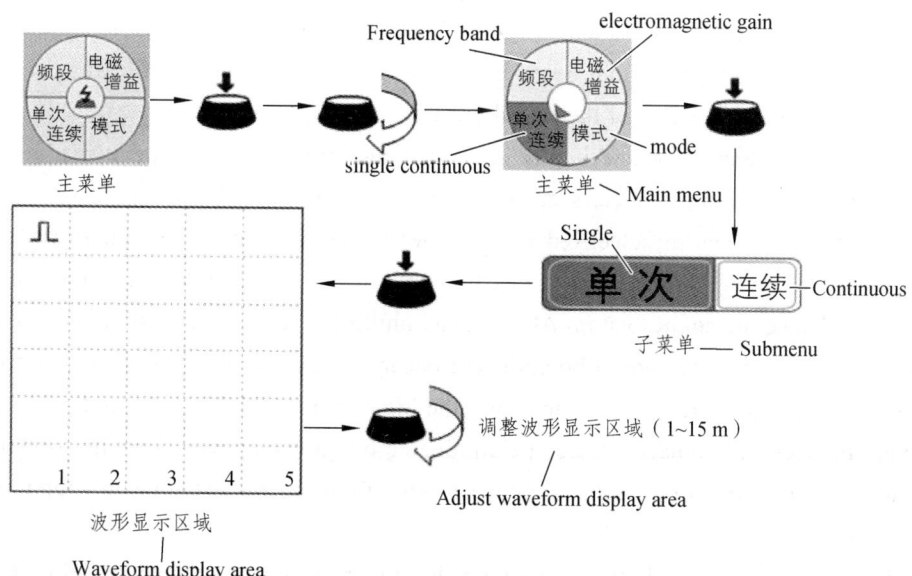

Figure 3-78 Specific operation process of single mode

(Ⅳ) Cable Fault Testing Method

1. Acoustic-Magnetic Fix Point Mode

(1) Open the instrument and enter the acoustic-magnetic fix point mode. Select the continuous mode. When fixing the point, first determine the approximate range of the fault (using the host for distance measurement), and then accurately locate within this range. During the fixing point process, initially, mark points can be made every 4–5 meters. When a regular "pa-pa" vibration sound is heard (the discharge sound at the fault point should be synchronized with the received electromagnetic wave, and the process of listening to the sound should refer to the received electromagnetic wave), slow down steps (every 1 meter) to locate the point.

At the same time, if a regular "pa-pa" vibration sound is heard (synchronized with the spark discharge sound in the air gap), but the distance displayed is 25.0 meters, it indicates that the fault point is too far from the probe or the vibration wave is too weak. In this case, continue searching forward. The maximum display range of the instrument is set to 25 meters. Because when the range is too large, the frequency of interference entering will increase, and the displayed erroneous data will also increase, leading to potential misjudgment by the testing personnel. Additionally, underground sound waves do not propagate too far, so a larger display range is meaningless.

(2) When approaching the ignition point and the distance displayed is within 15 meters, observe if the waveform area repeatedly shows waveform information with similar characteristics. Once observed, the probe can be moved based on the displayed distance and characteristic waveform. If the firing frequency is too fast or the waveform information is not clear, entering the single mode allows for a detailed analysis of the waveform. In this mode, the waveform can be analyzed by flipping through pages and observing the distance axis.

(3) When the pickup is placed above the fault point, the synchronized distance displayed on the locator is the smallest, the sound heard is the loudest, the electromagnetic wave signal is the strongest, and the recorded value of the sound wave is the highest.

(4) Sometimes when the probe is placed at the same point, the instrument readings may be different, such as displaying 5 meters at one time and 3.6 meters at another. This is actually a normal phenomenon. When the electromagnetic wave opens the door, other sound waves may be received before receiving the discharge spark sound wave. Any sound wave received by the instrument will cause the count to stop. At this time, multiple measurements should be taken at the same location and more data should be collected because interfering sound waves do not always come in at the same time difference. Therefore, the highest frequency data should be taken as the correct data. Interference signals can also be eliminated by observing waveform information.

(5) When there is continuous interference in the environment, listening to sound should be given priority.

(6) When encountering relatively soft soil, the probe should be connected below the probe. When testing, insert the probe into the ground and apply a little force in the vertical direction. Do

not pry or rotate it forcefully to avoid damaging the probe.

Typical waveform during testing is shown in Figure 3-97.

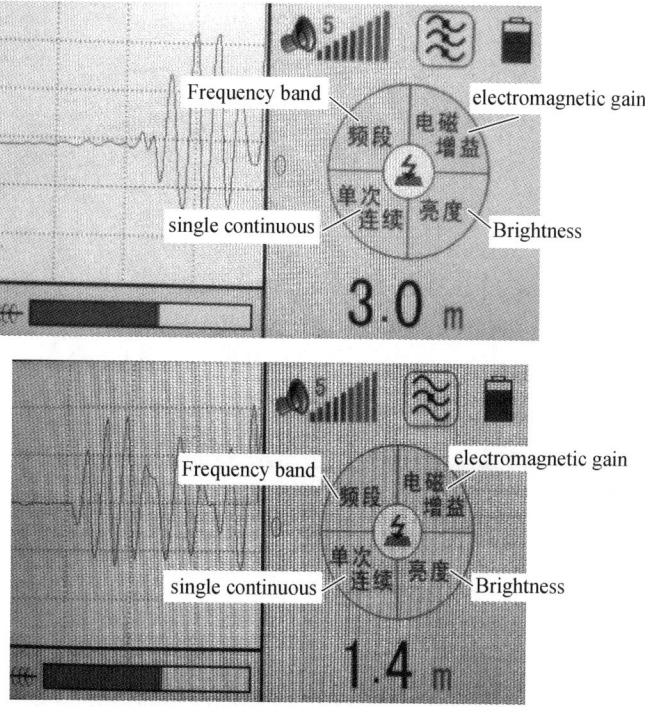

Figure 3-79　Typical waveform during testing

2. Step voltage mode

(1) Test principle

The power supply voltage for the current flowing into the soil from the fault point is negative polarity, and the surface potential of the soil is distributed in a funnel shape. The step voltage method precisely locates the lowest potential point or zero-crossing point of the step voltage in the soil using the probe, achieving precise positioning.

When there is a leakage current in the soil at the fault point during testing, the step voltage method can be used for precise positioning. Set the instrument to the step voltage mode, connect it to the A-frame, and conduct the test along the cable direction. After observing the regular waveform information, pay attention to the direction of the waveform's starting edge. When there is a sudden change in the direction of the starting edge, the fault point has been found.

When approaching the fault point, the potential difference will rapidly increase, and when measuring before and after the fault point, the potential difference reaches its maximum value. When the two electrodes are directly above the fault point and equidistant from it, the potential difference is zero, and the waveform amplitude is small, close to a straight line. When the two electrodes cross the fault point, the measured potential gradually decreases and the waveform's starting edge becomes opposite, with the amplitude gradually decreasing as well, as shown in Figure 3-80.

During the testing process, the gain adjustment is made based on the signal strength. The amplification gain can be directly adjusted by rotating the encoder keypad. Every time an adjustment is made, the waveform information in the waveform area will be automatically cleared.

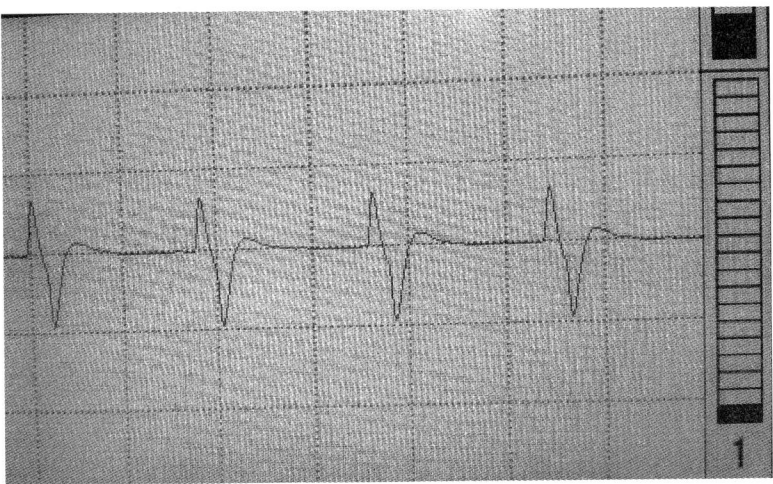

Figure 3-80　Waveform during the testing process

(2) Notes

① When the probe or the A-frame is first inserted into the ground, an unstable signal may be observed. Therefore, when observing the signal at a certain point, it is necessary to observe at least two to three cycles to determine a stable signal.

② If the signal amplitude is too large and reaches the maximum or minimum limit of the waveform display on the LCD, the gain should be reduced.

③ When the battery is low, please replace it with a new battery or charge it in a timely manner.

④ If there is a cement road surface or building above the cable where it is not possible to insert electrodes, move away from the cable and perform detection in a parallel direction.

⑤ When there are multiple grounding fault points, deal with one before searching for the next one.

Module 4　Training

Cable fault finding generally includes several steps, what are they?

_____
_____
_____

Worksheet 9 Insulating shoe and insulating glove test

Module 1 Operating Worksheet: Insulating shoe and insulating glove test

(Ⅰ) Test Name and Instrument	(Ⅱ) Test Objects
The withstand voltage test of insulating shoes and insulating gloves **The withstand voltage test instrument of insulating shoes and insulating gloves**	High-voltage insulating gloves, low-voltage electrical insulating gloves, high-voltage insulating shoes, low-voltage electrical insulating shoes, high-voltage rubber insulating boots, etc.
(Ⅲ) Test Purpose	(Ⅳ) Measurement Steps
The insulating boots and gloves are tested for voltage withstand to detect any defects or insulation hazards. This helps in preventing safety accidents from occurring	(1) Connect the output of the control box to the input of the test transformer. (2) Connect the instrument to the measurement device. (3) Connect the ground of the operation box to the high-voltage tail or ground of the high-voltage output. (4) Ground the grounding terminal of the control box. (5) Connect the high-voltage terminal to the insulating boot (glove) test vehicle's water resistance, and fill the water resistance with water. (6) Connect the signal line of the control box to the signal line terminal of the test vehicle, and ground the test vehicle. (7) Operate the withstand voltage test device of insulating boots (gloves) according to the operating procedure
(Ⅴ) Precautions	(Ⅵ) Technical Standards
(1) During the test, the operator should maintain a safe distance (less than 20 kV/m). (2) Before test, check the boots insulation resistance should be greater than 2 MΩ. (3) Check if the contacts of electrical components are loose. (4) Wipe the insulating support rods, electrodes, electrode rods, water tanks, and other parts with alcohol. (5) After the test, drain the water completely and dry all parts with a cotton cloth. Keep in a dry and well-ventilated place. (6) Storage areas should be free from gases, steam, chemical dust, and other explosive or corrosive substances that may significantly affect insulation. (7) Operated by professionals and strictly follow the operating procedures	DL/T 1476–2015 "Technical Regulations for Preventive Testing of Electric Safety Tools"

(Ⅶ) Result Judgment	(Ⅷ) Digital Resource
During the voltage withstand test, if the insulating cover, insulating gloves, or insulating boots are punctured, the current fluctuates abnormally, or the leakage current exceeds the standard, namely, leakage current of high-voltage insulating gloves exceeds 9 mA, and that of low-voltage insulating gloves exceeds 2.5 mA, and the leakage current of high-voltage rubber insulating boots exceeds 7.5 mA, they should be considered unqualified and prohibited from being used as insulation safety tools again	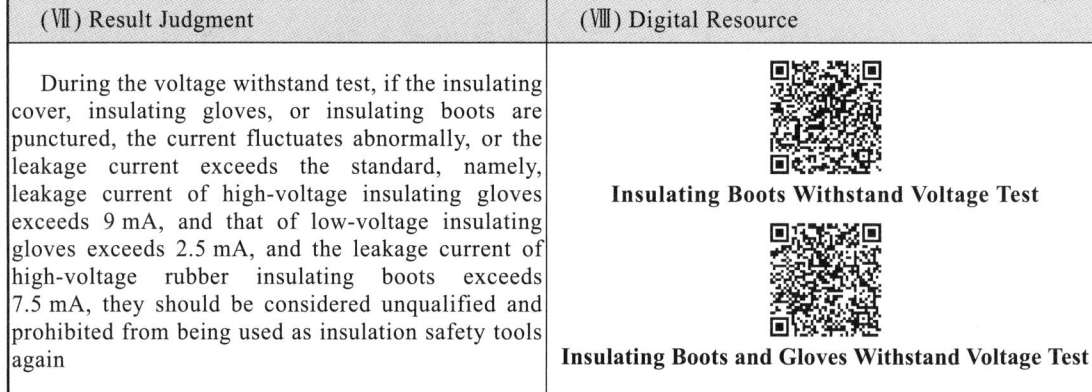 Insulating Boots Withstand Voltage Test Insulating Boots and Gloves Withstand Voltage Test

Module 2 Follow Me

Ⅰ. Insulating gloves and insulating boots

(Ⅰ) Insulating gloves

Insulating gloves, also known as high-voltage insulating gloves, are auxiliary safety tools used when operating high-voltage electrical equipment. They are used to operate high-voltage disconnect switches, high-voltage drop switches, install and remove grounding wires, and perform voltage testing on high-voltage circuits. When working on live low-voltage AC/DC circuits, insulating gloves are considered basic safety equipment, as shown in Figure 3-81.

Insulating gloves are commonly used safety tools and important insulation protection equipment in the maintenance, operation, and testing of electric power. There are two types of specifications for insulating gloves: 12 kV and 5 kV. The test

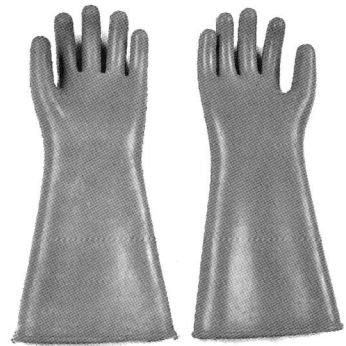

Figure 3-81 Insulating Gloves

voltage of the 12 kV insulating gloves is 12 kV. When working in high-voltage areas above 1 kV, they can only be used as auxiliary safety protective equipment and cannot touch live equipment. When working in areas below 1 kV, they can be used as basic safety equipment. That is, with the gloves on, both hands can touch live equipment below 1 kV (except for other parts of the body). 5 kV insulating gloves are suitable for general low-voltage electrical equipment. They are used as auxiliary safety equipment in voltage areas below 1 kV. In voltage areas below 250 V, they can be used as basic safety equipment. When working in voltage areas above 1 kV, the use of these insulating gloves is strictly prohibited.

(Ⅱ) Insulating Boots

The function of high-voltage insulating boots is to keep the human body insulating from the ground. They are auxiliary safety equipment used by personnel in high-voltage operations to maintain insulation from the ground. They can also be used as basic safety equipment to prevent step voltage and are important measures to prevent electric shock accidents, as shown in Figure 3-82. The power industry has special high-voltage insulating rubber boots with specifications of 20 kV, 25 kV, 35 kV, etc. In the workplace, insulating shoes should be selected according to the voltage level. Low-voltage insulated shoes are prohibited from being used as auxiliary safety equipment in high-voltage electrical equipment. High-voltage insulating boots can be used as auxiliary safety equipment in both high and low-voltage electrical equipment. However, whether wearing low or high-voltage insulating boots, direct contact with electrical equipment by hand is strictly prohibited.

Figure 3-82　Insulating Boots

(Ⅲ) Precautions for Using Insulating Gloves and Insulating Boots

(1) Before using insulating gloves and insulating boots, check if they have a valid certificate.

(2) Before using insulating gloves and insulating boots, conduct a visual inspection. They should not have any external injuries, cracks, bubbles, or burrs. If any issues are found, they should be replaced immediately. Insulating gloves should undergo an airtightness test. The specific method is to roll the gloves up from the cuff and pressurize the air slightly to the palm and fingertips to check for any air leakage. If leakage is detected, they should not be used.

(3) When using insulating gloves, pay attention to preventing sharp objects from piercing them.

(4) Before using insulating boots, check the voltage level of the workplace. It is prohibited to use insulating boots with withstand voltage lower than the required level in high-voltage electrical equipment. Insulating boots can only be used as auxiliary safety equipment on electrical equipment, and it is not allowed to directly touch electrical equipment with bare hands while wearing insulating gloves or boots.

(5) During the use of insulating boots, prevent sharp objects from piercing them.

(6) After using insulating gloves and insulated boots, store them in a dry place and avoid contact with oil and corrosive substances.

(7) Insulating gloves and insulating boots should be inspected every six months, and those that fail the withstand voltage test should not be used.

Ⅱ. Test Method for Insulating Gloves and Insulating Boots' Withstand Voltage

(Ⅰ) Insulating Gloves

(1) Place water with a resistivity not greater than 100 Ω·m inside the gloves. Use a metal ball with a diameter not exceeding 4 mm or tap water to fill the gloves. Then immerse the gloves in a metal basin filled with water, ensuring that the water levels inside and outside the gloves are at the same height. There should be a 90 mm exposed water surface portion on the gloves, which should remain dry during the test. The test connection is shown in Figure 3-83.

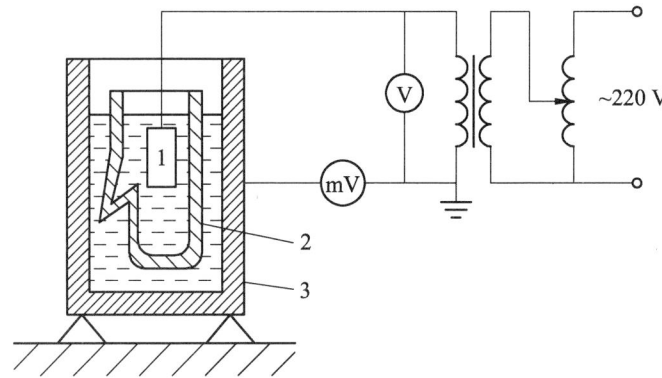

1–Electrodes; 2–Test sample; 3–Metal utensils for holding water.
Figure 3-83 Wiring schematic diagram of insulating glove withstand voltage test

(2) Apply a steady and uniform voltage increase to the specified voltage value of 8 kV. Maintain this voltage for 1 minute. Electrical breakdown should not occur during the test. If the measured leakage current is less than 9 mA, the gloves are considered to have passed the withstand voltage test.

(Ⅱ) Insulating Boots

(1) Place a metal plate with the same shoe size as the test item inside the boot as the inner electrode. Then, place a metal ball with a diameter not exceeding 4 mm on the metal plate, with a height of no less than 15 mm. Weld a copper plate with a diameter larger than 4 mm as an external connection wire, burying it inside the metal ball. The outer electrode is a water-soaked sponge placed inside a metal container. The test circuit diagram is shown in Figure 3-84.

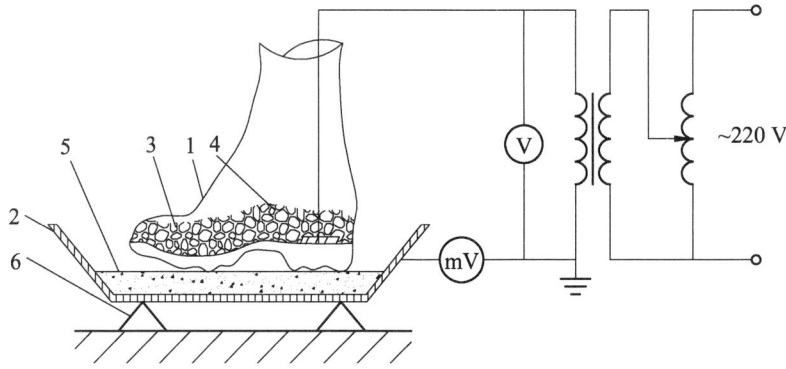

1–Tested boots; 2–Metal plate; 3–Metal ball; 4–Metal sheet; 5–Sponge and water; 6–Insulation support.
Figure 3-84 Wiring Schematic Diagram of Insulating Boots Withstand Voltage Test

(2) Apply a voltage increase at a rate of 1 kV/s from zero to approximately 75% of the specified voltage value (around 19 kV), and then increase the voltage at a rate of 100 V/s until it reaches 25 kV. Once the voltage reaches the specified value of 25 kV, maintain it for 1 minute. Electrical breakdown should not occur during the test. If the measured leakage current is less than 10 mA, the boots are considered to have passed the withstand voltage test.

III. Test Items, Periods, and Requirements (see Table 3-27)

Table 3-27 Withstand Voltage Test Standards of Insulating Gloves and Insulating Boots

Item	Period	Requirements			Description
Power-frequency Withstand Voltage Test	Half a year	Power-frequency withstand voltage/kV	Duration/min	Leakage Current/mA	Insulation resistance measurement is not required before and after the withstand voltage test
		Insulating gloves 8	1	≤9	
		Insulating boots 25	1	≤10	

IV. Precautions

During the test, the operator should maintain a safe distance (less than 20 kV per meter in air). The power frequency withstand voltage test bench must be reliably grounded, with a ground resistance of less than 0.1 Ω. Test the insulation resistance, which should be greater than 2 MΩ. Before use, check if the contacts of electrical components are loose, if the connections are good, and if all protection systems are functioning properly. Wipe the insulating support rods, electrodes, electrode rods, water tanks, and other parts with alcohol. After the test, drain the water completely and dry all parts with a cotton cloth. If not in use for a long time, store the water tank, electrode rods, and insulating support rods in a dry and well-ventilated place. The working and storage areas should be free from gases, steam, chemical dust, and other explosive or corrosive substances that may significantly affect insulation. It must be operated by professionals and strictly follow the operating procedures. Disconnect the power when removing the insulating boots (gloves).

(1) When performing the water immersion test for insulating boots (gloves), remove the sponge.

(2) The water level inside and outside the insulating boots (gloves) should be the same height, with 90 mm of water surface exposed. Ensure that the exposed portion of the insulating boots (gloves) is dry and clean. Then, place the high-voltage electrode inside the insulating boots (gloves) and clamp them securely.

(3) When performing the steel ball test inside the insulating boots, fill the water tank with water to fully soak the sponge. Place a metal plate of the same size as the boot size inside the boot. Place the high-voltage electrode inside the insulating boot, making contact with the metal plate. Then, put metal balls with a diameter not exceeding 4 mm on the metal plate, with a height of no less than 15 mm.

(4) Set up the testing site according to relevant regulations and ensure proper equipment connections. There should be dedicated personnel responsible for safety present to provide guidance.

(5) The grounding ends of the control box, test transformer, and test electrodes must be reliably grounded.

Module 3 Workshop

I. Device Structure and Installation

The insulating glove (boot) testing platform adopts a multiplexing structure, capable of performing tests on six pairs of insulated gloves or boots. The device platform consists of a bottom tug wheel, a flume ground electrode, a sponge ground electrode, a high-voltage test rectangular electrode strip, and a high-voltage wireless micro-ammeter.

1. Installation of Insulating Boots

Ensure that the withstands voltage test device of insulation safety equipment is in a power-off state, and that the testing platform is grounded. Place the sponge ground electrode on the flume ground electrode and pour in an appropriate amount of water to wet the sponge. Place the insulating boots between the sponge ground electrode and the high-voltage test rectangular electrode strip. Pour a box of test steel balls into the insulating boots, install the wireless milliampere meter on the high-voltage test rectangular electrode strip, and attach the test conductive chain. Then, hang the conductive chain into the insulating boots, ensuring that the ball of the test conductive chain makes contact with the test steel balls. Use the buoyancy of the sponge to support the weight of the insulated boots and steel balls.

2. Installation of Insulating Gloves

Ensure that the withstands voltage test device of insulation safety equipment is in a power-off state, and that the testing platform is grounded. Remove the sponge ground electrode from the flume ground electrode, and pour an appropriate amount of water into the tank. Hang the insulated gloves between the water tank ground electrode and the high-voltage test rectangular electrode strip using clamping bars. Pour the required amount of water into the insulating gloves according to the regulations. Install the wireless milliampere meter on the high-voltage test rectangular electrode strip and attach the test conductive chain. Then, hang the conductive chain into the insulating gloves, ensuring that the ball of the test conductive chain makes contact with the test steel balls.

3. Installation and Use of Intelligent Wireless Milliampere Meter

Intelligent wireless milliampere meter is shown in Figure 3-85.

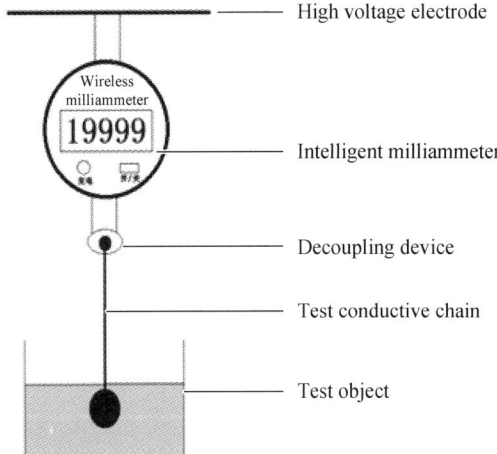

Figure 3-85　Intelligent wireless milliam meter

Ⅱ. Operational Steps

(Ⅰ) Steps

First, ensure that the withstands voltage test device of insulation safety equipment is in a power-off state, and that the test object has been sufficiently discharged before proceeding with the following operations:

(1) Enter the high-voltage test area and securely attach the top fixing nut of the intelligent wireless milliampere meter to the high-voltage end bolt of the rectangular test frame.

(2) Connect the high-voltage lead to the high-voltage end bolt, ensuring that the high-voltage lead is suspended and not in contact with the ground.

(3) Turn on the power switch of the intelligent high-flow meter and align the center of the matching test conductive chain with the suction cup at the other end. The test conductive chain is attached to the suction cup. Install insulating gloves (or boots).

(4) Exit the high-voltage test area and perform the power frequency withstand voltage test. Gradually increase the high voltage until it reaches the test voltage of the specimen. At this point, the digital display on the intelligent high-flow meter shows the leakage current of the specimen. The LCD display of the insulation safety tool's withstand test device allows real-time reading of battery power, leakage current, and the maximum breakdown current value of the specimen.

(5) During the process of increasing the voltage, consider the rated current value of the specimen or a leakage current value greater than the maximum specified in the specimen's specifications to set the protection current value. When the intelligent high-flow meter measures a leakage current value greater than the protection current value, the test conductive chain automatically disconnects, and the high voltage is interrupted. This indicates that the specimen is not qualified.

(6) During operation, the human body must not touch the instrument or high-voltage leads and maintain a safe distance.

(7) After completing the measurement test, turn off the power of the insulation safety tool withstand test device and fully discharge the high-voltage end and the specimen using a discharge rod.

(8) When discharging the high-voltage end using a discharge rod, first perform the discharge operation at the voltage equalization ball of the high-voltage test equipment, and then proceed with the discharge operation on the specimen. This is to prevent the large current impulse from affecting the intelligent high-flow meter during the discharge.

(9) Turn off the power switch of the intelligent high-flow meter, remove the high-voltage leads, unscrew the instrument, and store it properly.

(Ⅱ) Connect the wires properly according to the test requirements

Schematic diagram of test wiring is shown in Figure 3-86.

1. Power wiring for the control box

Connect to AC 220 V power supply.

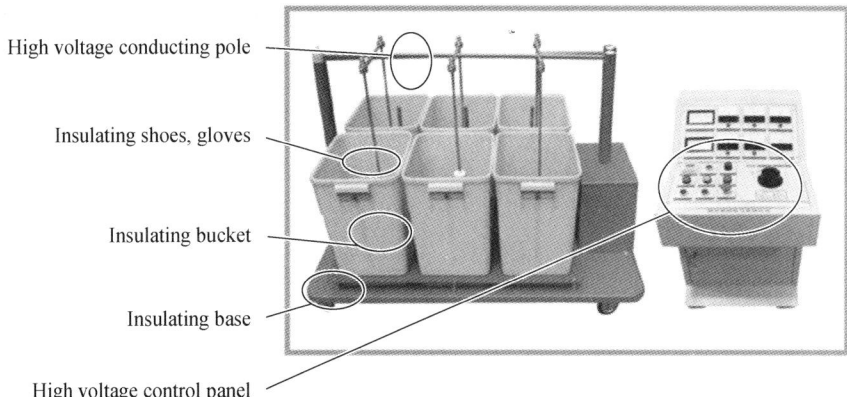

Figure 3-86　Schematic diagram of test wiring

2. Connection wires for the control box

One set of 2-core wires connects from the output of the control panel to the input of the test transformer. The other set of two wires consists of a green wire and a black wire, which correspond to the terminals between the control panel and the transformer, respectively for voltage-voltage (green wire) and ground-ground (black wire).

3. The ground wire for the transformer

The ground wire for the transformer must be grounded. Use a thin black wire with a clip on one end to connect to the ground terminal of the transformer and the other end to a reliable earth ground.

(Ⅲ) Control Box Panel Structure Diagram

Control box panel structure diagram is shown in Figure 3-87.

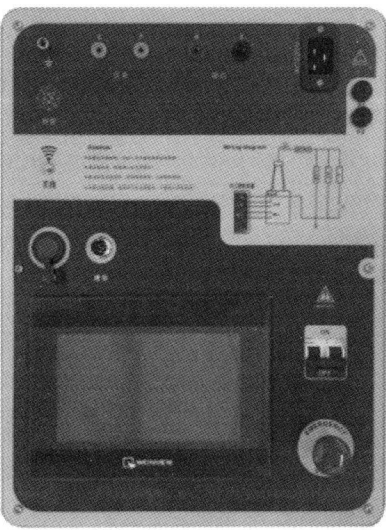

Figure 3-87　Panel structure diagram

(1) Antenna interface: Connects to the antenna to enable data collection of high-voltage leakage current of the isolation device.

(2) USB interface: Used to connect an external USB flash drive for exporting test data.

(3) Touch screen: 7-inch color touch screen that enables all human-machine interaction functions of the device.

(4) Buttons: Consists of five buttons for timing, start, stop, voltage increase, and voltage decrease, enabling partial human-machine interaction functions of the device.

(5) Power switch: Working power supply with a power indicator light.

(6) Communication interface: Intelligent high-flow meter gateway interface, optional function.

(Ⅳ) **Software Usage**

1. Initialization

After booting up, the device automatically performs system initialization.

2. Test Standby

After completing the boot-up initialization, enter the test standby interface. Based on the four circular buttons on the standby interface, access functions such as automatic endurance testing, manual testing, historical data respectively, and system settings, as shown in Figure 3-88.

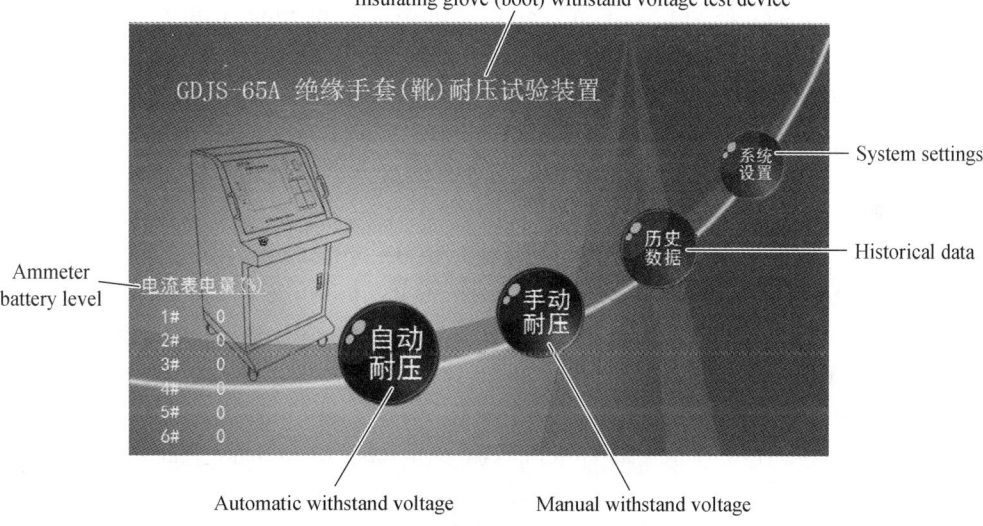

Figure 3-88 Test main interface

3. Automatic withstand voltage Testing

(1) After clicking on the automatic withstand voltage testing button, upon entering the main interface of automatic withstand voltage testing, the device automatically returns to zero, as shown in Figure 3-89.

(2) Real-time Acquisition Display Area: Under automatic withstand voltage testing, real-time monitoring of low voltage, low current, high voltage leakage current, and high voltage.

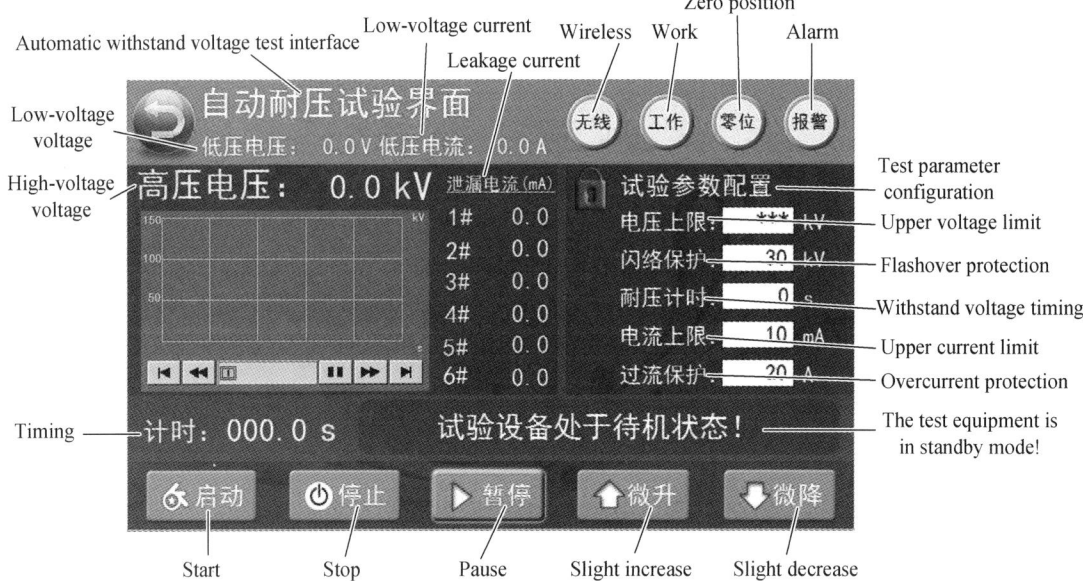

Figure 3-89　Automatic Withstand Voltage Testing Interface

(3) Test Parameter Configuration Area: This area has a parameter configuration protection lock. Click on the lock icon. After the lock icon turns green to indicate an unlocked state, configure parameters such as voltage upper limit, flashover protection, withstand voltage timing, current upper limit (with intelligent high-flow meter for leakage current), and overcurrent protection. The explanations for these parameter settings are as follows:

① The voltage upper limit is the withstand voltage value targeted during automatic boosting.

② Flashover protection refers to the critical protection value against high-voltage flashovers in automatic mode.

③ Withstand voltage time refers to the duration of the withstand voltage process.

④ The current upper limit refers to the current at which the intelligent high-flow meter disconnects. If the high-voltage current exceeds the current upper limit, it will be disconnected and regarded as insulation breakdown protection for insulating gloves (boots). After closing the lock, the setting value of the current upper limit will be wirelessly transmitted, and wirelessly received by the intelligent high-flow meter, and the screen will flash to indicate that the data has been received.

⑤ Overcurrent protection is the upper limit of the peak value of low-voltage current. If the low-voltage current exceeds the overcurrent protection threshold, it will be regarded as breakdown and protected.

(4) Information Display Zone 1: Wireless, Operation, Zero position, and Alarm four indicator lights. Their indications are as follows:

① Wireless light: Indicates the connection status between the control console and the high-voltage disconnection device gateway. If it is not illuminated, it indicates that the communication is not normal. If it is illuminated, it indicates that the communication is normal.

② Operation light: Indicates the operating status of the device. If it is not illuminated, it indicates that the test has not started. If it is illuminated, it indicates that the test is in progress.

③ Zero position light: Indicates the zero position output status of the regulator of the device. If it is not illuminated, it indicates that it is not in the zero position. If it is illuminated, it indicates that it is in the zero position.

④ Alarm light: Indicates the activation of the test alarm function. When the test starts, the buzzer will sound continuously for 2 seconds, and the alarm light will remain illuminated for 2 seconds before entering the withstand voltage test stage.

(5) Information Display Zone 2: Displays the test status and prompt information during the test process.

(6) Test Operation Zone: Consists of five touch buttons: Start, Stop, Pause/Resume, Fine Increase, and Fine Decrease.

4. Manual Withstand Voltage Test

(1) After clicking the Manual Withstand Voltage button, the device will enter the main interface of the manual withstand voltage test, and the device will automatically return to zero, as shown in Figure 3-90.

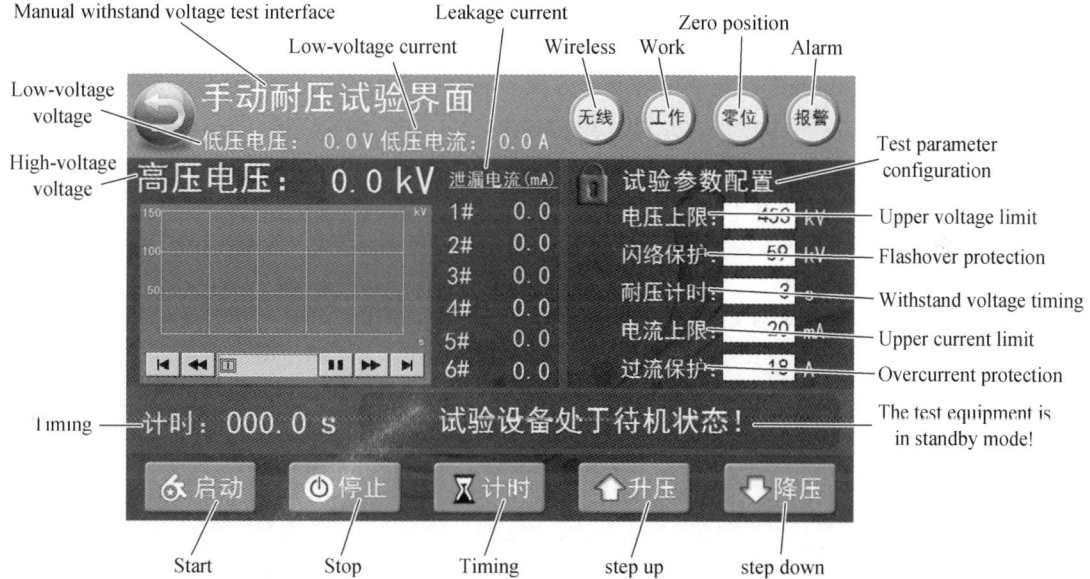

Figure 3-90　Manual Withstand Voltage Test Interface

(2) Real-time Acquisition Display Zone: In the manual withstand voltage test mode, it displays real-time monitoring of low voltage voltage, low voltage current, high voltage leakage current, and high voltage.

(3) Test Parameter Configuration Zone: This zone has a parameter configuration protection lock. Click on the lock icon, and once the lock icon turns green to indicate the unlocked state, proceed with setting the voltage upper limit, voltage ramp-up speed, withstand voltage timing, the

current upper limit (optional for intelligent high-flow meter disconnection current), and overcurrent protection parameters. The explanations for these parameter settings are as follows:

① The voltage upper limit is the withstand voltage value targeted during manual boosting.

② Flashover protection refers to the critical protection value against high-voltage flashovers in manual mode.

③ Withstand voltage time refers to the duration of the withstand voltage process.

④ The current upper limit refers to the current at which the intelligent high-flow meter disconnects. If the high-voltage current exceeds the current upper limit, it will be disconnected and regarded as insulation breakdown protection for insulating gloves (boots). After closing the lock, the setting value of the current upper limit will be wirelessly transmitted, and wirelessly received by the intelligent high-flow meter, and the screen will flash to indicate that the data has been received.

⑤ Overcurrent protection is the upper limit of the peak value of low-voltage current. If the low-voltage current exceeds the overcurrent protection threshold, it will be regarded as breakdown and protected.

(4) Information Display Zone 1: Wireless, Operation, Zero position, and Alarm four indicator lights. Their indications are as follows:

① Wireless light: Indicates the connection status between the control console and the high-voltage disconnection device gateway. If it is not illuminated, it indicates that the communication is not normal. If it is illuminated, it indicates that the communication is normal.

② Operation light: Indicates the operating status of the device. If it is not illuminated, it indicates that the test has not started. If it is illuminated, it indicates that the test is in progress.

③ Zero position light: Indicates the zero position output status of the regulator of the device. If it is not illuminated, it indicates that it is not in the zero position. If it is illuminated, it indicates that it is in the zero position.

④ Alarm light: Indicates the activation of the test alarm function. When the test starts, the buzzer will sound continuously for 2 seconds, and the alarm light will remain illuminated for 2 seconds before entering the withstand voltage test stage.

(5) Information Display Zone 2: Displays the test status and prompt information during the test process.

(6) Test Operation Zone: Consists of five touch buttons: Start, Stop, Pause/Resume, Slight Increase, and Slight Decrease.

(Ⅴ) Withstand Voltage Test Result

After the withstand voltage test is completed, it will automatically switch to the withstand voltage test result interface, displaying the parameter configuration, test result, and test status information. Please refer to Figure 3-91 for the automatic withstand voltage test result.

Ⅲ. Results Analysis

During the withstand voltage test, when the insulating gloves or insulating boots are punctured, the current exhibits abnormal fluctuations or the leakage current exceeds the standard values shown in Table 3-28, it should be considered as non-compliant. Specifically, if the leakage current of high-voltage insulating gloves exceeds 9 mA, the leakage current of low-voltage insulating gloves exceeds 2.5 mA, and the leakage current of high-voltage rubber insulating boots exceeds 7.5 mA, they should be deemed unacceptable and prohibited from further use as insulation safety tools.

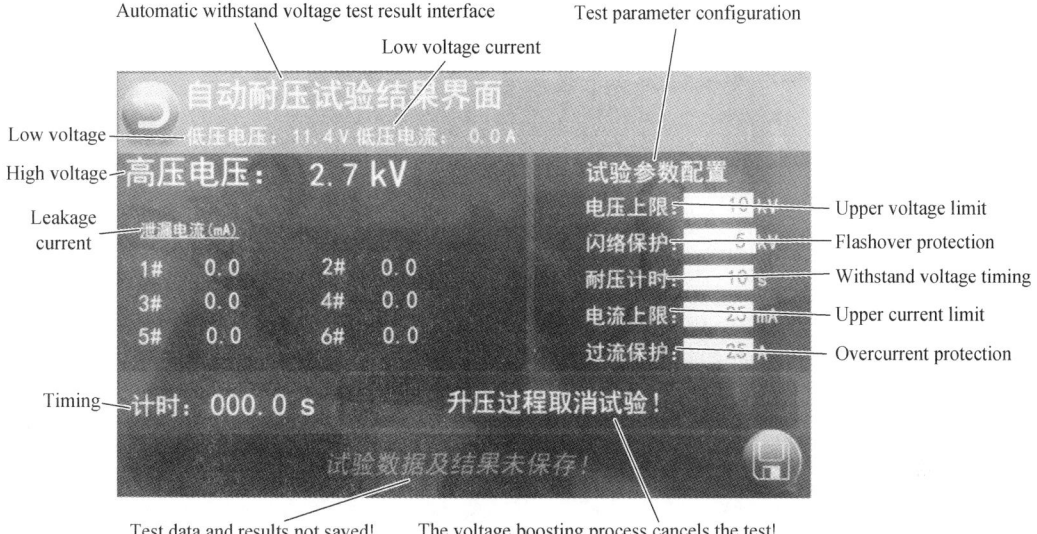

Figure 3-91　Automatic withstand voltage test result interface

Table 3-28　Insulation Tools Test Standards

ID	Name	Voltage Level /kV	Cycle	AC withstand voltage /kV	Time /min	Leakage Current /mA
1	Insulating Rod	6–10	Once per year	44	1	
		35–154		Four times the phase voltage		
		220		Three times the line voltage		
2	Insulating Glove	High Voltage	Once every six months	8	1	≤9
		Low Voltage		2.51		≤2.5
3	Rubber Insulating Boot	High Voltage	Once every six months	15	1	≤7.5
4	Insulating clamp	35 and below	Once per year	Three times the line voltage	5	
		110		260		
		220		440		

Module 4　Training

Please describe the test procedure for the insulating glove test.

_____
_____
_____

Worksheet 10　High Voltage Nuclear Phase Test

Module 1　Operating Worksheet: High voltage nuclear phase test

(Ⅰ) Security Tools	(Ⅱ) Safety Regulations
Insulating gloves, insulating boots, safety helmets, insulating mats, insulating rods, insulating pliers, electric test pens, grounding wires, shields, and signage	(1) Power off, conduct voltage testing, and hang the grounding wire before testing. (2) Take personal protective measures and environmental precautions
(Ⅲ) Preparations before testing	(Ⅳ) Testing Steps
(1) Follow the "two-wearing and three-carrying" principle. (2) Check if the high-voltage nuclear phase instruments are in good condition. (3) Record temperature and humidity	(1) Power off, check the power and hang the grounding wire before the test. (2) Wiring. (3) Record data. (4) Clean up the site
(Ⅴ) Precautions	(Ⅵ) Technical Standard
(1) During on-site testing, operations should be conducted according to the safety distance standards for high-voltage testing set by the power department. (2) The standard configuration for insulating rods is 3 meters, which corresponds to voltage levels of ≤ 220 kV. If measuring line voltages higher than 220 kV, please use insulating rods longer than 3 meters. (3) When performing nuclear phase operations, do not hold the insulation rod beyond the handle position	Based on DL/T 971–2005/IEC 61481:2004 "Portable nuclear phase instrument for Live Working in AC 1 kV–35 kV"
(Ⅶ) Result Judgment	(Ⅷ) Digital Resources
The nuclear phase is based on the X detector, with the A phase fixed as the reference. If the phase angle difference between the two detectors is within the range of -30° to 30° (-30° to 0° corresponds to 330° to 360°), and the Y detector detects the A phase, it is classified as "in-phase"; if the phase angle difference between the two detectors is within the range of 90° to 150° or 210° to 270°, it is classified as "out-of-phase". At the same time, the host's voice prompt is "in-phase" or "out-of-phase". When the phase angle difference is between 90° to 150°, the Y detector detects the B phase, which is a positive sequence; when the phase angle difference is between 210° to 270°, the Y detector detects the C phase, which is a reverse sequence	 **High Voltage Cable Nuclear Phase Test**

Module 2 Follow Me

Ⅰ. Overview of nuclear phase

For substations, power lines, and other installations where new laying, rewiring, or cable joints are performed, it is necessary to conduct a nuclear phase based on the phase markings in the power system to ensure a consistent phase. If the phases do not match, connecting two power sources through power lines with mismatched phases can result in the inability of the power grid to operate in a closed loop.

The high-voltage wireless digital nuclear phase instrument is used for phase calibration and phase sequence calibration of power lines and substations. It has the functions of phase verification, phase sequence measurement, and electrical inspection. It can verify the phase from 200 V to 220 kV automatically, such as 400 V, 10 kV, 35 kV, 66 kV, 110 kV, and 220 kV different grades of voltage transmission lines live working.

Ⅱ. The principle of the nuclear phase instrument

The principle of the nuclear phase instrument is to collect voltage signals from the power grid that have a phase relationship, modulate them, and compare the phase differences to determine whether they are in-phase or out-of-phase. The high-voltage nuclear phase instrument mainly consists of three parts: two sampling transmitters and a receiving display host. Nuclear phase instrument consisting of dual transmitters and a single host is shown in Figure 3-92. High voltage wireless nuclear phase instrument with dual transmitters and dual mainframes is shown in Figure 3-93.

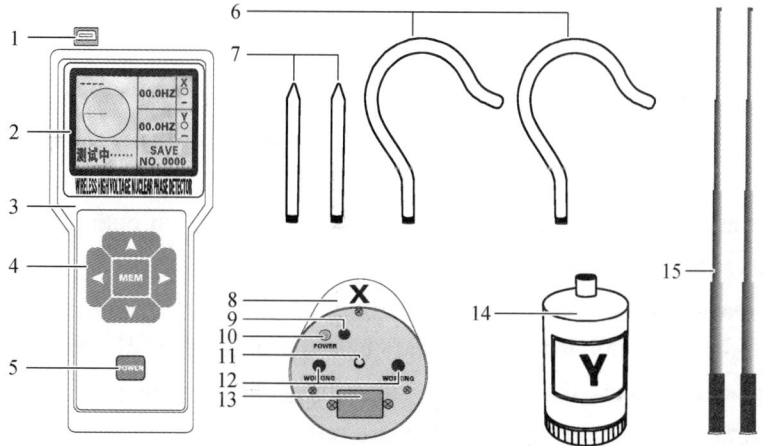

1–USB data download interface; 2–3.5-inch color LCD screen; 3–mainframe; 4–up, down, left, and right arrow keys and MEM control keys; 5–Mainframe power button (on/off); 6–Detector hooks (2 pcs); 7–Detector probes (2 pcs); 8–X-detector; 9–Power indicator light; 10–Detector POWER key (on/off); 11–Connection port for detector insulating rod; 12–Signal working indicator light; 13–Detector Battery Bottom Cover; 14–Y Detector; 15–Telescopic insulating rod (2 pcs).

Figure 3-92 Nuclear phase instrument consisting of dual transmitters and a single host

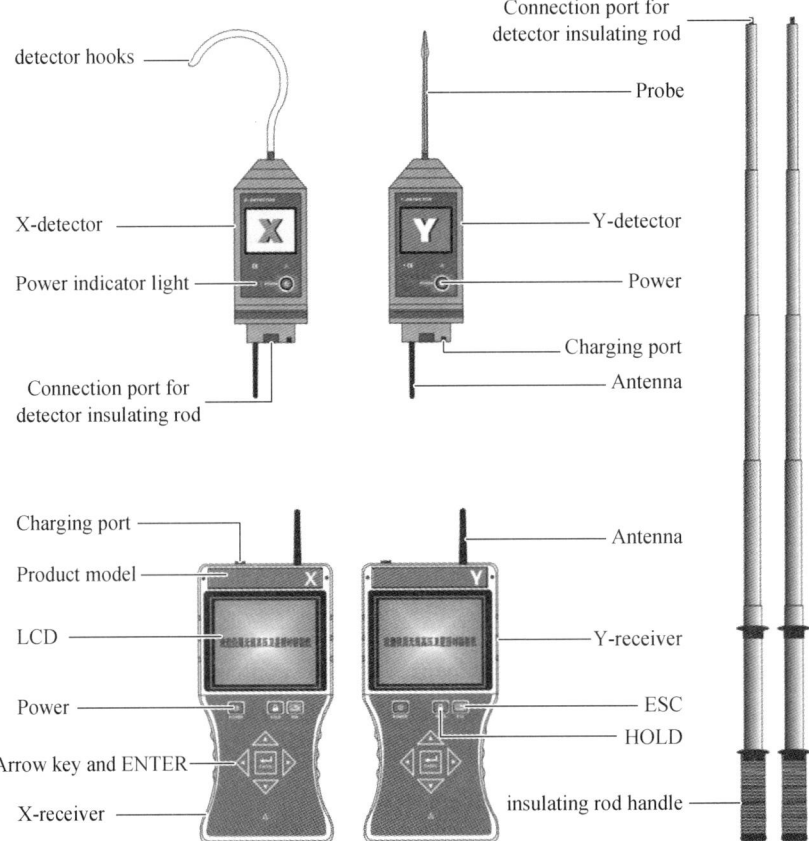

Figure 3-93　High Voltage Wireless nuclear phase instrument with Dual Transmitters and Dual Mainframes

During the test, the X emitter and the Y emitter are disconnected from each other electrically and are connected to the two phases for voltage phase sampling, and the voltage phase signal is transmitted wirelessly. First, it is judged whether the line is charged, and then the signal about the conductor is sent. The display receiver sends the digital signal to the receiving end through the wireless transmission module, receives the wireless signal of the two sampling transmission modules at the same time, calculates the voltage phase difference on both sides, emits the voice signal, and displays the phase result.

After the transmitting device works normally, the collected grid voltage signal with phase relationship is processed and modulated, and then transmitted to the receiving device. The receiving device demodulates the received grid voltage signal with phase characteristics and compares it with the grid voltage signal with phase relationship collected by the receiving device itself in real-time, and the phase difference value can be measured. The transmitter sends the phase and frequency signals of their respective lines back to the receiving host. The receiving host calculates the phase difference between the three lines, which has voice prompt functions such as "X signal is normal, Y signal is normal, in-phase, and out-of-phase". The high-voltage phase detector considers that the phase difference value is less than 0°–30° as the same phase, and the phase difference value is greater than 30° as the different phase.

The nuclear phase is based on the X detector, and the A phase is fixedly displayed. If the phase angle difference between the two detectors is in the range of -30°–30° (330°–360° is -30°–0°), the detection result of the Y detector is the A phase, which is the same phase qualitatively; if the phase angle difference between the two detectors is in the range of 90°–150° or 210°–270°, it is different phase. At the same time, the host voice prompts "same phase" or "different phase". When the phase angle difference is 90°–150°, the detection result of the Y detector is B phase, that is, sequential phase sequence. The phase angle difference is 210°–270°, and the detection result of the Y detector is C phase, namely inverse phase sequence.

Module 3 Workshop

Ⅰ. Connection Wiring

When the phase is checked on the spot, the nuclear phase instrument is detected on the same power grid first. One person operates and one person monitors. During the operation, the two transmitters are hung on the same conductor of the power grid first. When the two transmitters are working normally, the green light is on. Under normal working conditions, the receiver will display the phase angle and frequency of the two line voltages, and display the line voltage waveform, and the nuclear phase instrument is normal.

When operating the nuclear phase instrument on-site, one person operates and another person monitors. Operates, and records according to the operation steps gradually. The two transmitter devices are hung on the two conductors of the measured power grid first, and the display screen shows the phase difference and line frequency of the two lines. According to the above operation, the phase of the two power grids is determined phase by phase.

The nuclear phase can be done through contact or non-contact methods. Contact nuclear phase can be directly conducted on bare conductors below 35 kV and conductors with safety insulation skin below 110 kV. Non-contact phase comparison is used for bare conductors above 35 kV and lines above 110 kV. Regardless of the method, insulating rods must be used during operation.

The bare wire detector below 35 kV can contact the nuclear phase directly, and the bare wire above 35 kV adopts the non-contact nuclear phase. The non-contact nuclear phase is to gradually approach the detector to the measured wire. When the electric field signal is induced, the nuclear phase can be completed, so that there is no need to contact the high-voltage wire directly, which is safer.

Ⅱ. Testing Method

Connect the insulating rod, and power on the device. If the communication between the host and the detector is normal, the corresponding indicator light will illuminate, and the host will prompt "X signal normal" and "Y signal normal". If the communication is not normal, the indicator light will not illuminate.

For the nuclear phase, first, bring the X detector close to or contact with any phase line, and then bring the Y detector close to or contact with the other phase lines to be compared. During the high-voltage nuclear phase, the detector does not need to directly contact the high-voltage conductor. Gradually bring the detection hook of the detector close to the conductor, and when an electric field is detected, the detector will emit a "du-du-du" sound and the indicator light will continuously flash, indicating that the power testing function is completed, as shown in Figure 3-94. For low-voltage nuclear phase (400 V and below), especially for low-voltage distribution boxes, replace the metal detection hook with a metal probe.

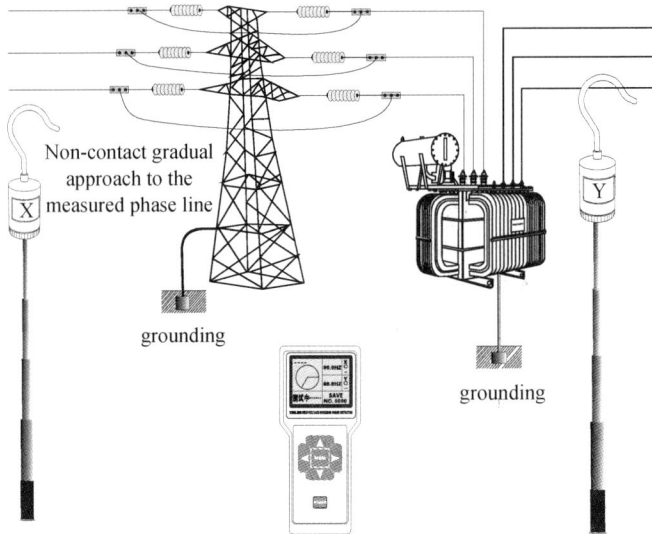

Figure 3-94   Nuclear phase test operation

During the non-contact nuclear phase, if the phase lines are relatively close to each other, choose a position that is far away from other conductors for testing.

Special attention: During the test, it is strictly forbidden to simultaneously hook onto two bare conductors, as it can cause a short circuit between the two conductors, which is extremely dangerous. Refer to Figure 3-95. Example of on-site nuclear phase operations is shown in Table 3-29.

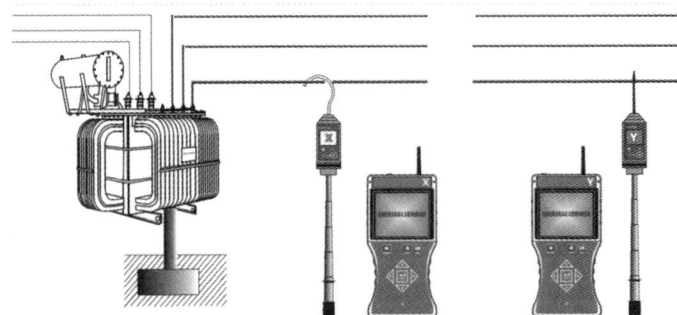

Figure 3-95   Misoperation: Strictly prohibited to simultaneously hook two bare conductors during the nuclear phase

Table 3-29 Examples of on-site nuclear phase operations

Nuclear phase for Bare Conductors Above 35 kV	The detector is connected to an insulating rod, and the rod is fully extended, so there is no need to install a detection hook or probe. The detector should be gradually brought close to the conductor for the non-contact nuclear phase. During the non-contact nuclear phase, the detector should try to avoid other conductors
Nuclear phase for Lines Below 35 kV	The detector can be connected to an insulating rod, and the rod is fully extended so that the detector can be hung on the line for contact nuclear phase
Nuclear phase for 380 V/220 V Low Voltage Municipal Power Supply Lines	The front end of the detector can be in contact with the live line for the nuclear phase, and there is no need to install a detection hook or probe. The height of the insulating rod should be adjusted according to the height of the line above the ground
Nuclear phase for Lines Below 100 V	Probe or detection hook can directly contact the conductor for the nuclear phase. If the voltage is too low, the auxiliary test line plug should be inserted into the charging port of the detector, and the other end of the auxiliary test line should be clamped to the grounding terminal or the cabinet door
Nuclear phase for High Voltage Switchgear with Live Indicator	The detector does not need to be connected to an insulating rod. Install the probe and insert it into the live indicator for the nuclear phase. If the voltage is too low, the auxiliary test line plug should be inserted into the charging port of the detector, and the other end of the auxiliary test line should be clamped to the grounding terminal or the cabinet door (This method is for the secondary side nuclear phase. The correctness of the nuclear phase result depends on whether the correspondences between L1, L2, L3, and the busbar are correct)
Nuclear phase for PT and CT Secondary Side Connection Points of Switchgear	The detector does not need to be connected to an insulating rod. Install the probe and insert it into the live indicator for the nuclear phase. If the voltage is too low, the auxiliary test line plug should be inserted into the charging port of the detector, and the other end of the auxiliary test line should be clamped to the grounding terminal or the cabinet door (This method is for the secondary side nuclear phase. The correctness of the nuclear phase result depends on whether the correspondences between L1, L2, L3, and the busbar are correct)
Nuclear phase for T-coupling of 10 kV/35 kV Enclosed High Voltage Cabinet	For an XY detector, it should be connected to an insulated rod and equipped with a detection hook. The detector can contact the T-head for the nuclear phase, and in general, there is usually no need to use a detection hook to contact for nuclear phase
Nuclear Phase for Five-Protection Switchgear	The detector should not be connected to an insulating rod, and there is no need to install a probe or detection hook. First, power off the busbar or move the trolley out of the tested switchgear. Then attach the detector to the busbar or trolley busbar, and use a strap to secure the detector tightly to the busbar or busbar. Turn on the detector, and then power on the switchgear for the nuclear phase
Nuclear phase for Primary and Secondary Sides of 10 kV/35 kV Transformer	The nuclear phase X detector should be connected to an insulating rod and detection hook, hanging on the primary circuit (10 kV/35 kV end) of the 10 kV/35 kV transformer. Y detector should be connected to an insulating rod and detection hook, hanging on the secondary circuit (400 V end) of the transformer

Module 4 Training

What are the project requirements of the high-voltage nuclear phase?

_____
_____
_____

Worksheet 11 Grounding Resistance Measurement of Ground Grid

Module 1 Operating Worksheet: Grounding Resistance Measurement

(Ⅰ) Test Name and Instrument	(Ⅱ) Test Objects
Grounding resistance measurement of ground grid **Ground Resistance Tester**	Measure the power frequency grounding impedance, contact voltage, step voltage, and soil resistivity of various grounding devices in power, railway, communication, mining, and other industries
(Ⅲ) Test Purpose	(Ⅳ) Measurement steps
To determine whether the ground resistance is within a reasonable range, thereby achieving the effectiveness of protecting the grounding device	(1) Disconnect the connection points between the main grounding conductor and the grounding electrode, as well as the grounding branch. (2) The voltage probe is inserted into the underground 20 m away from the grounding electrode, and the current probe is inserted into the underground 40 m away from the grounding electrode, both of them should be inserted vertically into the ground at a depth of about 400 mm. (3) Place the grounding resistance Tester on a level place near the grounding electrode and then wire it: ① Connect the E to and the measured grounding electrode with the shortest connection. ② Connect the P terminal to the voltage electrode 20 m away with a longer wire. ③ Connect the C terminal to the current electrode 40 m away with the longest wire. (3) Adjust the coarse adjustment knob according to the resistance requirements of the tested grounding electrode. (5) Multiply the reading on the fine adjustment dial by the multiple set by the coarse adjustment knob, and the result will be the measured ground resistance value of the grounding electrode

(Ⅴ) Precautions	(Ⅵ) Technical Standards
(1) In the straight-line extension directions of the tested electrode and the auxiliary electrode, keep a distance from underground conductors such as pipelines and water channels to avoid measurement errors caused by uneven soil resistivity. When it is not possible to stay far away from these underground conductors, try to make the distances between them and the grounding conductor intersect perpendicularly, without being parallel or overlapping. Do not have a large area of conductor on the ground between the three electrodes. (2) The total resistance of the extended test line should be less than 1 Ω, and the resistance under the use of humidity should be used to calibrate the data. (3) The soil moisture content, temperature, and additional salt will affect the earth's resistivity directly, thus changing the conductivity of the soil around the grounding resistance. (4) Ensure the conditions of no water on the ground, no rain for three days, and air humidity less than 90%. (5) Noise will cause interference in the measurement, resulting in inaccurate measurement results	(1) In accordance with GB 50057–2010 "Code for Lightning Protection Design of Buildings". (2) In accordance with JGJ/T 16–2008 "Code for Electrical Design of Civil Buildings"
(Ⅶ) Result Judgment	(Ⅷ) Digital resources
(1) For AC working grounding, the grounding resistance should not be greater than 4 Ω. (2) For safe working grounding, the grounding resistance should not be greater than 4 Ω. (3) For DC working grounding, the grounding resistance should be determined according to the specific requirements of the computer system. (4) The grounding resistance of the lightning protection grounding should not be greater than 10 Ω. (5) For shielding systems that use joint grounding, the grounding resistance should not be greater than 1 Ω	**Lightning Rod Grounding System Test**

Module 2 Follow Me

Ⅰ. Overview of Grounding Resistance

Grounding resistance refers to the resistance value between the grounding electrode, equipment enclosure, or building grounding electrode and the earth. The smaller the grounding resistance, the better it can divert leakage currents or lightning signals into the ground, preventing personal and equipment damage. Grounding resistance includes contact resistance between electrical equipment and grounding wires, resistance of grounding wires or grounding bodies themselves, contact resistance between grounding bodies and the earth, and resistance of the earth.

At present, the grounding resistance of the grounding grid is mainly measured by the power frequency large current three-pole method in the power system. In order to prevent the power frequency interference generated during the operation of the power grid and improve the accuracy of the measurement results, the insulation preventive test procedure stipulates that the test current of the power frequency large current method shall not be less than 30 A.

The 50 Hz data obtained from measurements in strong interference environments of substations are particularly suitable for environments with large fluctuations in railway substation return current. The anti-interference grounding grid grounding resistance meter is a specialized instrument used to measure the grounding resistance and conductivity of a grounding grid between grounding points. It is capable of measuring both grounding impedance and resistance simultaneously, providing an accurate reflection of the grounding grid's actual characteristics.

II. Principles of Grounding Resistance Measurement

Grounding resistance refers to the resistance displayed when releasing current through a grounding device, numerically equal to the ratio of the voltage drop across the grounding device to the current flowing into the earth.

$$R = \frac{U}{I} \tag{3-1}$$

In the equation, the U represents the voltage to the ground of the grounding device, which is the potential difference between the grounding electrode and the reference point at zero potential with respect to the earth. The I represents the current flowing into the earth through the grounding device.

Therefore, the principle of measuring grounding resistance is based on equation (3-1): applying a current I to the grounding electrode or grounding grid, measuring the voltage U on the grounding electrode (grid), and dividing the voltage by the current to obtain the grounding resistance.

III. Five Common Connection Methods for Grounding Resistance Testers

The grounding resistance tester is specifically used to test the electrical integrity of grounding devices. Similar to an insulator fault detector, it is part of electrical testing and a regular inspection item. The main objects of measurement include the grounding resistance of power line towers, transformers, etc. The tester commonly utilizes the following five connection methods:

(I) Clamp Measurement Method

This method measures the grounding resistance at each grounding point in a multi-point grounding system without disconnecting the grounding connections to prevent potential hazards. It is suitable for multi-point grounding systems where the connections cannot be disconnected and the resistance of each grounding point needs to be measured.

(II) Double-Clamp Method

This method measures the grounding resistance of a single grounding point in a multi-point grounding system without using an auxiliary grounding rod. When connecting, the current clamp is connected to the corresponding socket, and both clamps are clamped on the grounding semiconductor. The distance between the two clamps should be greater than 0.25 m.

(III) Two-Pole Method

There must be a known well-grounded point, and the measured result is the sum of the resistances between the tested ground and the known ground. If the resistance of the known ground is much smaller than that of the tested ground, the measurement result can be used as the result of the tested ground. It is suitable for areas with dense buildings or closed areas such as cement floors where it is impossible to hit a grounding rod.

(Ⅳ) Three-Pole Method

There must be two grounding rods, one auxiliary ground, and one detection electrode, and the distance between each grounding electrode must be no less than 20 m. It is mainly used for foundation grounding, construction site grounding, and lightning protection ball lightning rod grounding.

The measurement connection methods of the grounding resistance tester for grounding grids can be divided into the three-pole method and the four-pole method. The schematic diagrams are shown in Figures 3-96 and 3-97.

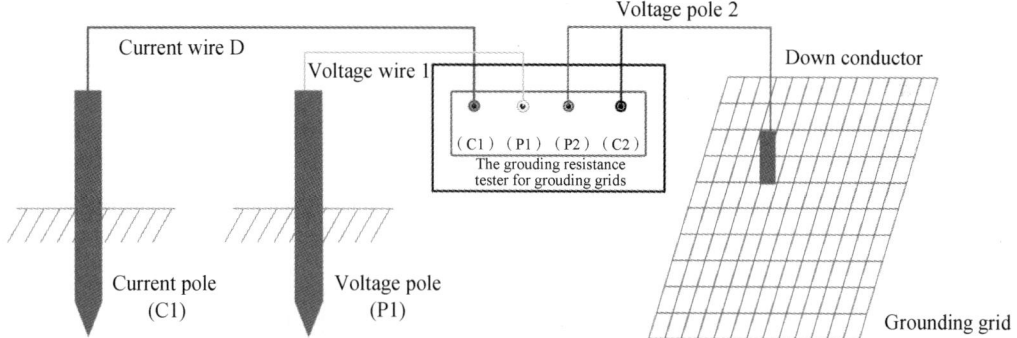

Figure 3-96   Three-Pole method wiring diagram for grounding resistance measurement

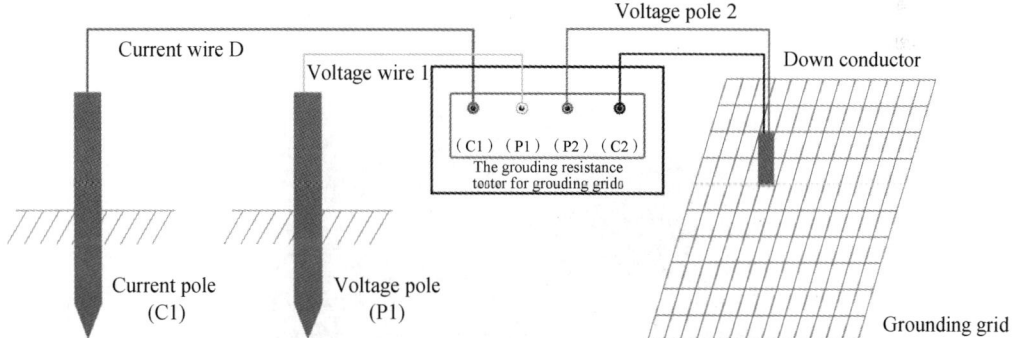

Figure 3-97   Four-pole method wiring diagram for grounding resistance measurement

During the three-pole method connection:

Measurement of current line D: The wire diameter is ≥ 1.5 mm^2, and the length is 3 to 5 times the diagonal length of the ground grid.

Measurement of voltage line 1: The wire diameter is ≥ 1.0 mm^2, and the length is $0.618D$.

Measurement of voltage line 2: Connect to the ground grid being tested.

Measurement wire: Connect to the ground grid being tested.

(Ⅴ) Four-Pole Method

Similar to the three-pole method, it replaces the three-pole method in low grounding resistance measurement and eliminates the influence of measurement cable resistance on the measurement results. This method is a more accurate grounding resistance measurement method, as shown in Figure 3-97.

During the four-pole method connection, two connecting wires are drawn from the grounding rod of the grounding grid and connected to the voltage terminal P2 and the grounding grid terminal C2 of the instrument. Following the measurement operation steps, the instrument will automatically eliminate wiring errors during the four-pole method measurement.

In the three-pole straight-line method and the triangular method for measuring grounding resistance, it is necessary to temporarily create an auxiliary current electrode in the distance to provide a return path for the current.

Module 3 Workshop

Ⅰ. Instrument operation interface

Panel diagram is shown in Figure 3-98.

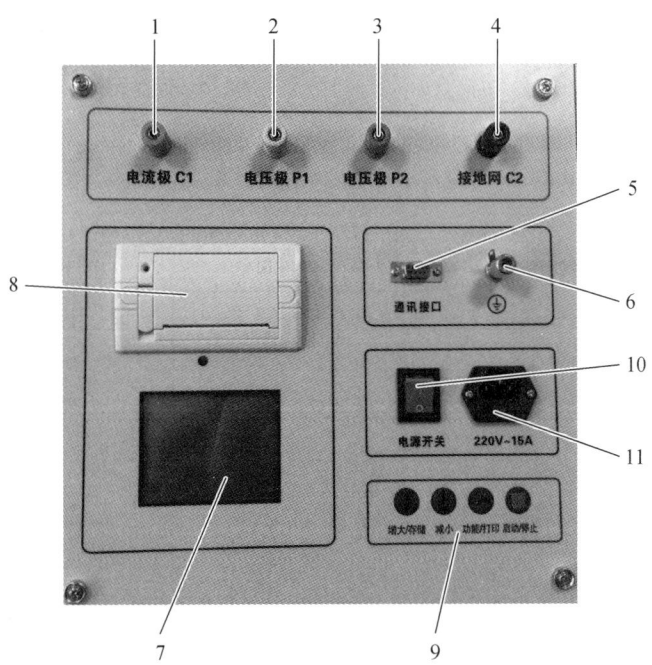

1–Current pole (C1); 2–Voltage pole (P1); 3–Voltage pole (P2); 4–Grounding grid (C2); 5–232 serial port; 6–Grounding post; 7–LCD screen; 8–Printer; 9–Keyboard; 10–Power switch; 11–220 V power outlet.

Figure 3-98 Panel Diagram

"▲" Increase/store key-modify the menu content, using circular scrolling.

"▼" Decrease key-modify the menu content, using circular scrolling.

"▶" Function / print key-Select the menu item, and the selected item is displayed in reverse white font.

"■" Start/Stop Key-Press this key on the [Test] option to enter the test state.

Ⅱ. Operation Steps

1. Use the three-pole method for wiring

(1) Use a multimeter to check if there is any open circuit in the test current line, voltage line, and grounding wire. Check if the rust on the grounding rod has been cleaned off and if the depth of its burial is appropriate (>0.5 meters).Also, check if the connection between the test wires and the grounding rod is conducting. If it is not conducting, please resolve the issue and reconnect.

(2) The length ratio between the test current line and the test voltage line should be 1∶0.618. The length of the test current line should be 3 to 5 times the diagonal length of the grounding grid.

(3) Connect one end of the test current line and the test voltage line to the instrument as required, and then lay them out in parallel. Connect the other end of each line to two grounding rods respectively.

(4) Double-check the laid-out test lines. Connect one end of the multimeter to the test current line or the test voltage line, and connect the other end to the grounding wire. If there is no resistance value displayed, it indicates an open circuit. Confirm that everything is in good order before proceeding with the test.

(5) After confirming that the connections are correct, connect the instrument to an AC 220 V/50 Hz power source and power it on.

(6) Select Grounding Resistance Test. On the startup interface, when moving the cursor to the Grounding Resistance Test, use the increase and decrease keys to modify the test wire length. In general, choose 20 meters as the test wire length, as shown in Figure 3-99.

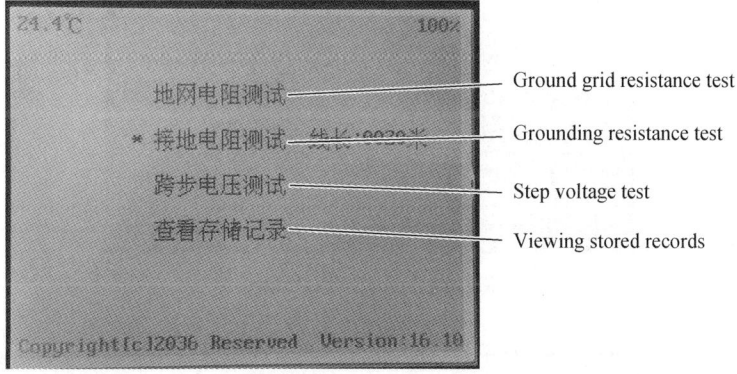

Figure 3-99　Moving the Cursor to Grounding Resistance Test

(7) Move the cursor to the desired test item, and press the Start/Stop key to enter the test for that item. Press the Measure key to start the measurement.

(8) After the instrument displays the end of the test, record the test data (multiple measurements can be repeated), as shown in Figure 3-100.

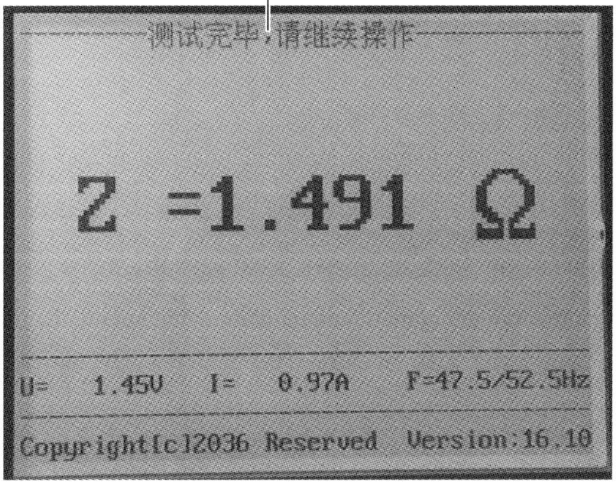

Figure 3-100   Test data

① The meanings of the measurement results are as follows:

Z=1.491 Ω: Grounding impedance value.

U=1.45 V: Test voltage value.

I=0.97 A: Test current value.

F=47.5/52.5 Hz: Represents the test frequency of (50 ± 2.50) Hz.

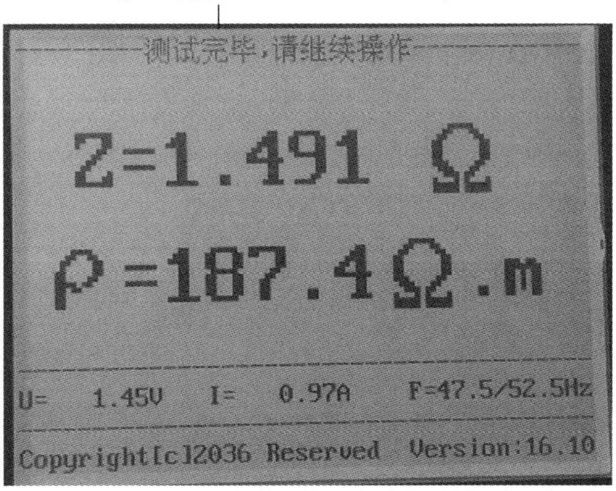

Figure 3-101   Test data

② The meanings of the measurement results are as follows:

Z=1.491 Ω: Grounding resistance value.

ρ=187.4 Ω·m: Soil resistivity.

U=1.45 V: Earth voltage value.

I=0.97 A: Test current value.

$F=47.5/52.5$ Hz: Represents the test frequency of (50 ± 2.50) Hz.

(9) After turning off the instrument's power, remove the connections, and the test process ends.

2. Double-clamp method for measuring grounding resistance value

The double-clamp method is suitable for measuring the grounding resistance value of an independent multi-point grounding system. As shown in the Figure 3-102, there is no need to hit a grounding stake to measure the grounding resistance value in a multi-point grounding system. Both current clamps A and B are clamped to the test grounding lead wire simultaneously. Note that the direction of the two current clamps should be the same, and the distance between them should be greater than 30 cm. The two current clamps must not be interchanged, otherwise, errors may occur.

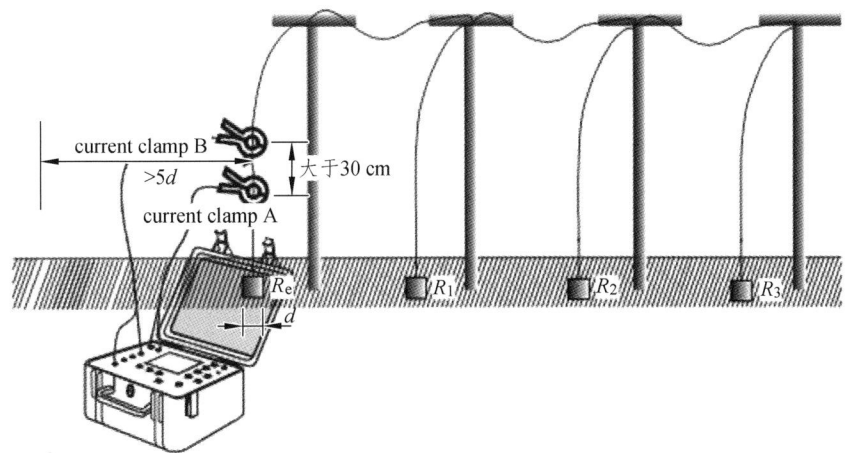

Figure 3-102　Wiring Diagram for Double-clamp Method of Grounding Resistance Measurement

Press the red "TEST" button on the instrument to start the test. After the test is completed, stable data will be displayed, which represents the grounding resistance value of the tested grounding electrode $R=R_e+R_1//R_2//R_3$, where $R_1//R_2//R_3$.

III. Precautions

When the test current is 0.0 A, it may be due to poor contact between the current wire connection and the current electrode grounding stake, or there are too few grounding stakes, which reduces the loop resistance. The depth of the grounding stake should not be less than 0.5 m, and the resistance of the current stake should be less than 80 Ω.

If the measurement value displayed by the instrument is extremely low (<0.01 Ω), it may be due to the voltage wire not being connected.

(1) Do not wind the leads.

(2) Keep the voltage wire as far away from the current wire as possible.

(3) Both sides of the grounding clamp should be tightly clamped to the test grounding lead wire to prevent poor contact caused by paint or rust.

(4) To prevent over current protection, select a current of 2 A.

To ensure smooth testing, please check the contact points between the test wires and the grounding stakes with a multimeter before testing, and check if there is any open circuit in the wires already laid.

Module 4 Training

What is the difference between the three-pole method and the four-pole method when conducting grounding resistance testing?

___
___
___

Evaluation

ID	Test 1	Test 2	Test 3	Test 4	Test 5	Test 6	Test 7	Test 8	Test 9	Test 10	Test 11
Project name	DC resistance test	Loop resistance measurement	Winding deformation test	Oil chromatography analysis	Insulation oil dielectric strength test	Switch characteristic test	Cable fault test	Withstand voltage test	Transformer ratio group test	Grounding resistance measurement of grounding grid	Three-phase nuclear phase test
Test equipment	DC resistance rapid tester	loop resistance tester	One winding deformation tester, one laptop for controlling the data collection of the tester, and three dedicated test cables	One set of oil chromatography analysis instruments	Insulation oil dielectric strength testing instrument	One set of multifunctional switch characteristic testing instruments	One set of cable fault comprehensive testing devices	One set of withstand voltage testing equipment	Multifunctional transformer ratio group testing instrument	One set of grounding resistance testing instruments	Wireless digital nuclear phase meter
Test content	Transformer DC resistance test	GIS switchgear loop resistance measurement	transformer winding deformation test	transformer oil chromatography analysis	transformer oil dielectric strength test	circuit breaker switch characteristic test	high-voltage cable fault test	insulating shoe and glove withstand voltage test	transformer ratio group test	grounding resistance measurement of grounding grid	power grid nuclear phase test
Project requirements	(1) Explain the principle of this high-voltage test project. (2) Conduct on-site demonstrations and explain the insulation structures and materials to be tested. (3) Follow the testing procedures for personal protection, ensure reliable test connections, good grounding, and proper implementation of operation and protective measures. (4) Complete the process of measurement. (5) Write the test report											
Step 1: Tool preparation	(1) Safety tools: voltage tester, insulating pole, grounding wire, connecting wires, discharge rod, warning fence, and warning signs. (2) Insulating tools: insulating gloves, insulating boots, safety helmet, safety belt, insulating mats, insulating ladder. (3) Tool inspection and arrangement: ① Divide the testing area and operation area, correctly place the testing equipment, and maintain a safe distance from the test object. ② Prepare relevant test cables, alligator clips, grounding wires, etc. ③ Prepare other tools: a multimeter, thermometer, humidity meter, etc											
Step 2: Risk Control	(1) Before the test, ensure proper "two-wearing and three-carrying" (wear working uniform, wear insulating boots, wear a safety helmet, wear insulating gloves, and use a voltage tester). (2) Set up barriers at the testing site and hang signs outside saying "No entry, high voltage danger". (3) Take necessary precautions for working at heights if required. (4) During the test, contact with the equipment is prohibited. (5) Prior to the test, the test personnel must follow the requirements of the "Work Permit" and take necessary safety measures											
Step 3: Test wiring	(1) Connect according to the requirements of the test project, and all test connections must be reliable and secure. (2) The testing instruments must be reliably grounded. (3) Sufficient safety testing distance should be maintained between personnel and equipment. (4) Establish and implement safety supervision and protection systems before and after the test											

Continued

ID	Test 1	Test 2	Test 3	Test 4	Test 5	Test 6	Test 7	Test 8	Test 9	Test 10	Test 11
Step 4: Test operation	(1) Electrical testing must be carried out by two or more persons, with one person supervising and another operating. (2) Connect the instrument to the corresponding current and voltage terminals, select the appropriate range, and then start testing. (3) After the test is completed and the data is read and confirmed, fully discharge and remove the test connections to end the measurement		(1) Electrical testing must be carried out by two or more persons, with one person supervising and another operating. (2) Before the test, the line ends of the transformer under test should be fully discharged. (3) Disconnect all leads of the transformer (including overhead lines, enclosed busbars, and cables), and keep the test leads away from the transformer bushing. (4) During the test, accurately record the positions of each tap switcher. For a no-load tap switcher, ensure that each measurement is taken at the same tap position for comparison.	(1) Take an oil sample and ensure that the sample is not contaminated. (2) Place the oil sample in the chromatograph. (3) Set the parameters of the chromatograph and then perform the measurement. (4) Record the chromatographic data and graphs. (5) Clean the testing instrument.	(1) Electrical testing must be carried out by two or more persons, with one person supervising and another operating. (2) Enter the instrument's operating interface to set the test parameters. (3) Ensure that the testing instrument is reliably grounded. (4) In the power-off state, when filling the oil sample into the oil cup, it should flow slowly along the inner wall of the cup to reduce air bubbles. During the operation, it is not allowed to touch the electrodes, the interior of the oil cup, or the oil sample with hands. After the oil cup is filled, it must be left to stand for 10–15 minutes before starting the voltage test.	(1) Electrical testing must be carried out by two or more persons, with one person supervising and another operating. (2) Follow the testing requirements and parameters for each test item (such as switch opening/closing time, in-phase and interphase synchronization, bounce time, voltage, etc.). (3) Set the test time and voltage parameters according to the corresponding test items. (4) Perform tests on switch characteristics.	(1) Electrical testing must be carried out by two or more persons, with one person supervising and another operating. (2) Use a megohmmeter to determine the type of cable fault (low-resistance fault, high-resistance fault, flashover fault, open circuit, etc.). (3) Set the cable measurement parameters (cable type, total length, etc.) based on the type of cable fault.	(1) Electrical testing must be carried out by two or more persons, with one person supervising and another operating. (2) Enter the instrument's operating interface to set the test parameters. (3) Voltage should be slowly increased from zero during the voltage rise, and the voltage rise speed should not be too fast to prevent sudden voltage increase. After the test is completed, the voltage should be lowered to zero. (4) The call and response system must be followed during high-voltage testing.	(1) Electrical testing must be carried out by two or more persons, with one person supervising and another operating. (2) When removing the transformer busbar or wire, fully discharge the high and low-voltage windings of the transformer. (3) Connect the instrument terminals to the transformer windings respectively, and ground the instrument grounding terminal.	(1) Connect according to the test requirements. (2) The connections should be firm and reliable. (3) Pay attention to the distance between the test lines E line, P line, and C line. (4) Select the appropriate range. (5) Separate the tested grounding electrode from other grounding bodies, and try to separate the measurement circuit from the power grid. (6) The tested grounding electrode should be disconnected from the ground wire.	(1) For high voltage testing, the insulating pole must be connected and fully extended, and the insulating handle end should be held by hand. (2) The safety withstand voltage level of this insulating pole is up to 220 kV. When the voltage exceeds 35 kV, a non-contact nuclear phase must be used. Direct contact with bare wires above 35 kV is strictly prohibited as it poses a risk of electric shock, which can cause personal injury or equipment damage. (3) It is forbidden to hook onto two bare wires at the same time, as this may cause a short circuit between the two wires. (4) Connect the insulating pole to the detector. For voltages below 400 V, use a metal probe for connection. For voltages above 400 V, use a metal hook for connection. When operating, always use the insulating pole.

Continued

ID	Test 1	Test 2	Test 3	Test 4	Test 5	Test 6	Test 7	Test 8	Test 9	Test 10	Test 11
Step 4: Test operation			(5) The transformer core must be reliably grounded with the enclosure, and the tester's enclosure and impedance measurement enclosure must be reliably grounded with the transformer enclosure. (6) Ensure that the connection clamps of the impedance measurement are tightly in contact with the bushing. If there is conductive paste or rust on the bushing, it must be wiped clean with sandpaper or a dry cotton cloth		(5) Pay attention to observing any changes in the test specimen during high-voltage testing. (6) If any abnormal conditions are discovered during the test, the test should be stopped to identify the cause	(5) Record data and analyze whether the data complies with the regulations. (6) Clean up the testing site	(4) Perform a rough measurement on the cable to determine the location of the cable fault, and use the high-voltage pulse method and precision measurement method for cable fault location. (5) Use the acoustic method to confirm the specific location of the cable fault. (6) Excavate to confirm the location of the cable fault	(5) If any abnormal conditions are discovered during the voltage rise test, the test should be stopped to identify the cause. (6) After changing the wiring or completing the test, the voltage should be lowered, the power should be disconnected, and the test equipment should be discharged by grounding. Only after confirming that the tested object is reliably grounded can the dismantling and installation of test lines be conducted in the test area	(4) Turn on the power supply, and set the wiring method, standard ratio, and tap switcher voltage ratio according to the transformer nameplate. (5) Select "start testing" and press "confirm" to display the group, ratio, and ratio error. Use the directional keys to switch to different tap positions for testing	(7) When measuring the ground resistance, insert the potential probe 20 m underground from the grounding electrode, and insert the current probe 40 m underground from the grounding electrode. (8) Turn on the power supply and start the test to display the ground resistance value	(5) According to the voice recording of the host, determine whether it is "in-phase" or "out-of-phase". (6) Verify with the phase identification of the power grid
Step 5: Data recording	(1) Record the name and model of the recording instrument equipment. (2) Record the name and model of the test sample. (3) Record the test data. (4) Record the historical test data of the test sample											
Step 6: Result analysis	Organize the site and confirm that it is clean after completion. (1) The comparison of the test data and the factory test report should comply with the test procedures. (2) The comparison of the test data and the handover test report should comply with the test procedures. (3) The test data should comply with the standards of the electric power industry or national test regulations											

Project 4

High Voltage New Technology

Worksheet 1 Hot Washing

Module 1 Operating Worksheet: Hot Washing

(Ⅰ) Test Name and Instrument	(Ⅱ) Test Objects
Water-making equipment, high voltage cleaning pump, cleaning water tank, cleaning spray gun, and water pipe	Insulating parts of contaminated electrical equipment
(Ⅲ) Test Purpose	(Ⅳ) Test Steps
(1) Remove various pollutants on various electrical equipment (dust, oil, moisture, salt, etc.) timely, prevent and eliminate pollution flashover and fog flashover of power equipment, and reduce and avoid the huge economic losses caused by power outages. (2) Improve the electrical insulation value of the equipment and mitigate secondary pollution. (3) Reduce the economic losses caused by power outage maintenance equipment greatly	(1) Make sure that the insulation of the equipment is good before the hot washing, and hot washing cannot be used for damaged and low-value insulators. (2) Prepare the insulating water needed for flushing. (3) Measured wind speed needs to meet operational requirements

Continued

(V) Precautions	(VI) Technical Standards
(1) When flushing insulators, attention should be paid to the wind direction. It is necessary to flush the downwind side first and then the upwind side. For the insulators arranged in the upper and lower layers, the lower layer should be flushed first and then the upper layer. Pay attention to the washing angle to prevent flashover of adjacent insulators in the sputtered water mist. (2) Different washing sequences will have different insulation surface states, which will affect the flashover voltage. One is to wash the outer layer from the bottom up or the wire side in order, and the splash wet surface is the smallest. other dirty layers are not wet. the other is to rinse from top to bottom, and the upper layer of the dirt has not been rinsed clean, and some of the insulation surface has been soaked	(1) TB/T 13395—2008 Guidelines for Hot Washing of Electric Power Equipment (2) DLT 1467—2005 Technical Specifications for Hot Washing Operation of 500 kV Transmission and Substation Equipment. (3) DLT 1468—2005 Electric Vehicle-mounted Hot Washing Device
(VII) Result Judgment	(VIII) Digital Resources
(1) Clean the dust on the equipment, and reduce the dust accumulation on the equipment. (2) improve the insulation level of the equipment effectively, and reduce the occurrence of discharge along the surface of the equipment. (3) Lightning arresters and poorly sealed equipment should not be water-washed	

Module 2　Follow Me

Ⅰ. Overview of hot washing

As atmospheric pollution will be attached to the magnetic column of the insulating porcelain vase, the rainy days are prone to pollution, which can lead to insulation breakdown seriously caused by short circuits or grounding accidents, so it is necessary to clean the insulating part of the substation equipment regularly in order to prevent the pollution of the equipment, to avoid the reduction of the insulation of the equipment, which will lead to the breakdown or short circuit disconnection accidents, such as shown in Figure 4-1.

Usually, the electrical equipment works in a strong electric load for a long time. The comprehensive pollutants caused by the environment, oxidation, and slow corrosion cause the contact resistance of the electrical equipment contacts to increase significantly, and the arcing is serious when the contacts act, resulting in high-temperature ablation, damage to the electrical appliances, and even accidents. Power equipment operating in a strong electromagnetic field, due to long-term outdoor exposure, has a strong affinity for various pollutants in the environment. Over time, a layer of dirt accumulates, posing the biggest risk of "pollution flashover" and "fog flashover." This leads to significant direct and indirect economic losses each year. Therefore, it is necessary to timely remove dust, oil stains, moisture, salt, and other pollutants from various electrical equipment such as overhead insulators and rod insulators of railway catenary systems, as shown in Figure 4-2, in order to reduce flashover faults and pollution flashovers of the support insulators of the catenary system.

Figure 4-1 Isolating switch in short-circuit disconnection accident

Figure 4-2 The support insulator of the OCS experiencing pollution flashover

  It is important to prioritize cleaning the insulators of the line in the substation, the porcelain bottles of the support columns for disconnecting switches, and the bushings of switch transformers in the substation. However, all conductive metal parts within the station, such as the power lines, transformer bodies, and knife switch contacts, should not be cleaned. It is also crucial to prevent water used for cleaning from entering the terminal boxes and avoid water ingress secondary wiring, as shown in Figure 4-3 and 4-4.

Figure 4-3　35 kV station transformer

Figure 4-4　The on-site layout of incoming lines at a substation

Figure 4-5 illustrates the on-site hot washing scene at a substation.

Figure 4-5 Substation hot washing scene site

II. The functions of hot washing

(1) Prevent and eliminate pollution flashovers and fog flashovers in power equipment, reducing and avoiding the significant economic losses caused by power outages.

(2) Help improve the electrical insulation value of the equipment and mitigate secondary pollution. It restores the normal surface impedance of circuit boards and components, providing special protection and ensuring that the equipment operates in optimal conditions.

(3) Reduce the contact resistance of electrical equipment contacts, reduce the power consumption of electrical equipment, and improve the working efficiency of equipment.

(4) Restore the normal heat dissipation capacity of electronic equipment, reduce the power consumption of electrical equipment, and improve the working efficiency of equipment.

(5) Reduce the economic losses caused by power outage maintenance equipment greatly.

Module 3 Workshop

I. Operating Procedures for Hot Washing

1. Basic Security Measures

Anyone entering the live cleaning and maintenance site should wear the prescribed working uniform and comply with the relevant safety regulations of the user (such as wearing a safety helmet with a face shield). Cleaning tools used should have insulating handles, and the exposed conductive parts should be insulated to prevent accidental short circuits during operation. During work, insulating shoes, cotton long-sleeved work clothes, and gloves should be worn, and the operation should be conducted while standing on an insulating mat or dry insulating material. Figure 4-6 shows personnel wearing insulating protective gear operating a handheld water gun.

Figure 4-6　Personnel wearing insulating protective gear operating a handheld water gun

2. Selection of Water Resistance Rate and Water Column

The pollution intensity is called salt density, which refers to the mass of pollutants such as cigarette ash, cement dust, and chemical materials that adhere to the insulation surface of electrical equipment per unit area due to external environmental pollution. It is used to measure the degree of insulation contamination, and its unit is mg/cm^2.

There is a difference in insulation between water used for hot washing and ordinary water. Common water contains impurities, and tap water and drinking water also contain various mineral ions, which can conduct electricity. However, the water used for hot washing is industrially filtered and has a very high resistivity, equivalent to an insulating medium. This is the safety standard for hot washing. The water used for hot washing cannot be drunk by humans.

Before the operation of hot washing, the contamination of the insulator should be mastered. When the salt density is greater than the critical salt density in Table 4-1, it is generally not suitable for hot washing, otherwise, the water resistance rate should be increased. hot washing is not suitable for arresters and poorly sealed equipment.

Table 4-5 Water Resistivity and Specific Creepage Distance

Item	Specific creepage distance of Substation post insulator/(mm/kV)							
	14.8–16 (normal type)				20–31 (anti-fouling type)			
Water resistivity /($\Omega \cdot$ cm)	1,500	3,000	10,000	50,000 and above	1,500	3,000	10,000	50,000 and above
Critical salt density/(mg/cm^2)	0.02	0.04	0.08	0.12	0.08	0.12	0.16	0.2

The resistivity of water used for hot washing is generally not less than 1,500 $\Omega \cdot$ cm. When flushing 220 kV substations, the resistivity of the water should not be lower than 3,000 $\Omega \cdot$cm, and it should comply with the requirements in Table 4-1. Before each flushing operation, the resistivity of the water should be measured using a qualified resistivity meter. Water samples should be taken at the water outlet for measurement. If water is collected in containers such as water trucks, the resistivity of the water should be measured for each batch. The resistivity of the water has a significant impact on ensuring the personal safety of the operators and the safety of the equipment. The resistivity of water has a significant impact on ensuring the personal safety of the operators and the safety of the equipment. This influence is mainly reflected in the insulation level between the water resistivity and the water column. Different heights of water columns and different water resistivities correspond to different power frequency discharge voltages. Figure 4-7 shows the relationship between water resistivity and its power frequency discharge voltage.

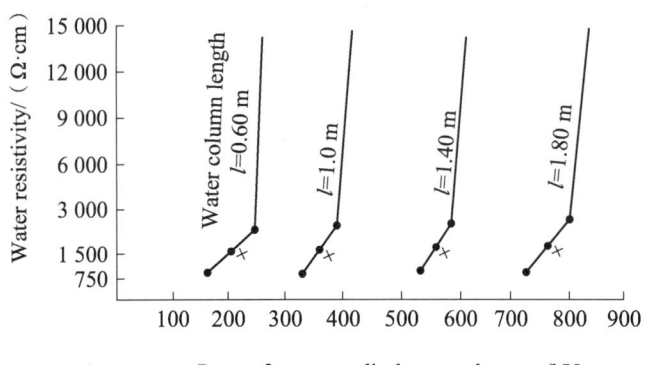

Figure 4-7 Relationship between water resistivity and power frequency discharge voltage

Tools for hot washing use water columns as the main insulation, while water pipes and insulation operating rods serve as auxiliary insulation. The water column is the key component that withstands voltage, but the insulation of the water column mainly depends on its length. At the same time, the diameter of the water gun nozzle also has a significant impact on the insulation.

Large, medium, and small hot washing tools with water columns as the main insulation (those with nozzle diameters of 3 mm and below are called small water washers; those with diameters of

4–8 mm are called medium water washers; those with diameters of 9 mm and above are called large water washers). The length of the water column between the water gun nozzle and the charged electrode must not be less than the requirements specified in Table 4-2. Both large and medium-sized water-flushing gun nozzles should be reliably grounded.

Table 4-2   Distance between the water gun nozzle and the charged electrode

Voltage level/kV	Nozzle diameter/mm			
	3 and below	4–8	9–12	13–18
63 (66) and below	0.8	2	4	6
110	1.2	3	5	7
220	1.8	4	6	8

3. Precautions for hot washing

Before conducting hot washing, it is necessary to ensure that the insulation of the equipment is in good condition. Generally, insulation with zero or low values and ceramic cracks should not be washed, as washing will cause changes in the insulation status on the surface of the insulator and along the potential gradient. Due to the reduced insulation performance and structural changes of insulators with low or zero values and ceramic cracks, the insulation changes at the ceramic crack under the action of a strong electric field will be more significant when immersed in water. Therefore, inspections should be conducted before washing to ensure that the equipment's insulation status is good. If the insulation performance of the insulator is found to be reduced, it should be replaced first before considering other work. If there is no reliable technical means of identification, damaged and low-value insulation should not be subjected to hot washing.

When washing insulators, attention should be paid to the wind direction. It is necessary to wash the downwind side first and then the upwind side. For insulators arranged in upper and lower layers, the lower layer should be washed first, followed by the upper layer. The washing angle should also be considered to prevent nearby insulators from experiencing flashover in the splashing water mist. Different washing sequences will result in different insulation surface conditions, which will affect the flashover voltage. One method is to wash the layers sequentially from bottom to top or from the conductor side outward, minimizing the wetting of the splashing surface and avoiding wetting of other layers of dirt. Another method is to wash from top to bottom, which may result in partial wetting of the insulation surface before the upper layer of dirt is completely washed off. Figure 4-8 shows a field where sequential flushing is performed.

Hot washing should generally be conducted in good weather conditions. It is not suitable to conduct hot washing when the wind speed exceeds 4 levels, the temperature is below 0 °C, during rainy, snowy, sandstorm, foggy, or thunderstorm weather. Therefore, there is no contamination of the water used for hot washing during operation. However, the water that has passed through the washing equipment may wash away the dirt on the surface of the insulation equipment, making the water no longer clean.

Figure 4-8　Wash the site in sequence

The safety distance between a person and a charged object must not be less than the regulations specified in Table 4-3.

Table 4-3　Voltage Levels and Human Safety Distances

Voltage level/kV	10	35	63 (66)	110	220	330	500
Safety distance/M	0.4	0.6	0.7	1.0	1.8 (1.6)	2.2	3.4 (3.2)

The above "safe distance" includes the minimum distance between the cleaning operator's metal tools (such as spray guns, etc.) and the energized equipment being cleaned. Therefore, during the hot washing process, both the supervisor and the operator must pay attention to their positions, as shown in Figure 4-9.

Figure 4-9　Positions of personnel at the washing site

Ⅱ. Water Washing Vehicle

Due to the exposure of insulators of electrified railway overhead contact lines to the external environment throughout the year, suspended particles, dust, smoke, and other external substances in the air can adhere to them, forming accumulated dirt. The dirt tightly bonds with the surface of the insulators over time, making it difficult to clean even with natural wind and rain. In humid weather, some salt compounds dissolve and transform into conductive water film solutions. The accumulated dirt on the surface of the insulators continues to absorb moisture in harsh weather conditions, leading to an increasing leakage current. This can cause pollution flashovers of the insulators under normal operating voltage, resulting in widespread and prolonged power failures that pose a serious threat to railway transportation safety.

Currently, the main method used in domestic electrified railway power supply sections to clean insulators is manual wiping during power outages of the overhead contact system. However, this method has drawbacks such as high labor intensity, low efficiency, and poor safety. Manual water washing by personnel also involves a heavy workload and high safety risks. In order to remove the accumulated dirt on the surface of contact system insulators, the railway power supply department has introduced water-washing vehicles for insulators. These vehicles effectively handle the contaminants on the insulators of the electrified railway overhead lines. Figure 4-10 shows an example of a contact system water washing vehicle, and Figure 4-11 depicts the on-site operation of the water washing vehicle for the contact system.

Figure 4-10 Overhead Contact System Water Washing Vehicle

Figure 4-11 On-site Operation of the Overhead Contact System Water Washing Vehicle

Dust can accumulate on the insulators of the contact system, causing them to become dirty. When ceramic insulators are heavily soiled, simple washing may not meet the cleanliness standards. For insulators that are out of reach on the outer side of the lines, a cloth is used to clean stubborn areas. The hot washing process begins at a speed of 5 kilometers per hour, targeting the insulators located 6 meters above the ground.

In the southern power grid, intelligent hot-washing operation vehicles have been deployed for on-site operations. Figure 4-12 shows an example of a robot-intelligent hot-washing operation vehicle.

Figure 4-12　Robot-intelligent hot-washing operation vehicle

Module 4　Training

Please explain the hot washing.

_____
_____
_____

Worksheet 2   High Voltage Live Detection

Module 1   Operating Worksheet: High Voltage Live Detection

(Ⅰ) Test Name and Instrument	(Ⅱ) Test Objects
Sensor, infrared detector	Transmission line, transformer, circuit breaker, disconnect switch, instrument transformer, bushing, lighting arrester, power capacitor, GIS, reactor, power cable
(Ⅲ) Test Purpose	(Ⅳ) Test Steps
(1) Timely detection of potential equipment defects. (2) Comprehensive monitoring of equipment operation. (3) Reducing the workload and hazards for maintenance personnel. (4) Lowering the operational and maintenance costs of equipment	(1) Take protective measures. (2) electroscope and confirm the charged equipment. (3) Wiring according to different test items. (4) Start the charged detection. One person operation and another person supervision. (5) Record data. (6) Organize instruments and clean up the site. (7) Quit
(Ⅴ) Precautions	(Ⅵ) Technical Standards
Regularly inspect the live detection system. Determine the monitoring cycle based on the equipment being tested	(1) GB 7597–2007 Sampling Methods for Electricity Oil (2) GB/T 7601–2007 Moisture determination method of transformer oil in operation (3) GB/T 14542–2017 Maintenance and management regulations of transformer oil in operation (4) DL/T 596–2021 Preventive test procedure for power equipment (5) DL/T 572–2021 Operation rules of power transformer (6) GB/T 7252–2021 Guide for analysis and determination of dissolved gases in transformer oil (7) GB/T 17623–2017 Gas Chromatographic Determination of Dissolved Gas Content in Insulating Oil
(Ⅶ) Result Judgment	(Ⅷ) Digital Resources
(1) Whether the high-voltage equipment is in normal condition. (2) Issuing warnings when the high-voltage equipment is abnormal	

Module 2   Follow Me

Ⅰ. Overview of Live Detection

    State monitoring is widely used in the power system, and electrical equipment testing is mainly divided into online monitoring and live detection. Online monitoring refers to the continuous and uninterrupted automatic detection of the electrical state of the live operation of electrical equipment through testing instruments while the equipment is in operation. It can detect potential equipment failures, especially in the status detection of intelligent substations, where it plays an extremely important role.

Live detection refers to the monitoring of equipment status on-site without a power outage. It is one of the effective measures for real-time and uninterrupted monitoring and discovering equipment defects. High-voltage live detection realizes the all-around monitoring of high-voltage equipment through sensors and data transmission technologies. The difference between live detection and power outage detection lies in whether power outage detection is used to confirm the electrical performance of the equipment. power outage detection is also called preventive testing of power equipment.

The objects of live detection include: transmission line, transformer, circuit breaker, disconnect switch, instrument transformer, bushing, lightning arrester, power capacitor, GIS, reactor, power cable, and other major primary equipment, as shown in Figure 4-13. It can detect the problems of primary electrical equipment in the operation state, including partial discharge, overheating faults, gas leakage faults, etc., to avoid major accidents.

Figure 4-13   Capacitor and reactor with compensation function in traction network

The significance of live detection is to detect the hidden defects that may exist in equipment during long-term operation while the equipment is running, which can be timely identified, treated according to the level and type of faults, and flexibly arrange maintenance cycles, reduce power outages, and improve equipment reliability.

II. The basic method of live detection

The current live detection technology mainly includes the transformer, GIS and SF_6 circuit breaker partial discharge test, transformer oil chromatography, capacitive equipment capacity and dielectric loss charged test, zinc oxide arrester leakage current, ultrasonic detection, ultraviolet acoustic imaging, infrared imaging detection method and so on.

1. Ultrasonic Method

When there are gaps, cracks, delaminations, interlayers, and other defects in the metal of electrical equipment such as transformers, high-voltage switchgear, GIS, etc., ultrasonic waves propagate to the interface between the metal and the defect, resulting in partial or complete reflection. These ultrasonic waves cannot be heard by humans. Ultrasonic testing is a method that utilizes instruments to receive ultrasonic signals that undergo refraction and reflection at the interface of insulating materials. After conversion and processing by electronic circuits, it can detect defects such as cracks and fractures in the insulating materials. This method has the function of relatively locating faults, and the changes in waveform characteristics correspond to the depth, position, and size of the defects.

The ultrasonic method can detect internal faults in equipment by emitting and receiving ultrasonic signals. It can also detect equipment faults by only receiving ultrasonic and electromagnetic wave information emitted during localized discharge faults inside the equipment. The frequency of ultrasonic waves is in the range of 20 to 100 kHz. The results are displayed in dB (mV) values, curves, or images, providing functions such as fault identification and localization. Acoustic-optic measurement combines acoustic measurement and fiber optic measurement. When partial discharge occurs inside a transformer, it generates ultrasonic waves that exert pressure on the installed optical fibers inside the transformer, causing deformation. This leads to changes in the length of the optical fibers and their refractive index. By demodulating the modulated waveform, the source and discharge location of the ultrasonic waves can be detected using a demodulator, providing accurate technical support for maintenance.

2. Infrared and ultraviolet imaging method

Thermal faults in power equipment can be divided into external thermal faults and internal thermal faults. External thermal faults include heating caused by resistance changes at wire connections, connectors, etc. Internal thermal faults include uneven distribution of thermal electric fields caused by heating at contact connection points of solid, liquid, and gas dielectrics inside the equipment.

Infrared imaging is a non-contact detection technique that allows the visualization of defects that are not visible to the naked eye. It can provide clear visual images of these defects, enabling real-time and online monitoring and diagnosis of most electrical equipment faults. In high-voltage measurement scenarios, infrared imaging is particularly sensitive to defects such as resistance losses, core losses, dielectric losses, temperature differences caused by uneven voltage distribution, and temperature differences caused by oil-immersed equipment with oil deficiency.

The principle of the infrared imaging method is that the temperature field of different parts of high voltage electrical equipment in operation can be judged accurately by using the infrared thermal imaging method and different colors to distinguish and display whether there is too high local temperature on the surface and inside of power equipment. To determine whether the equipment has insulation defects such as dielectric loss or resistance loss, it is particularly sensitive to early fault defects and the insulation status of electrical equipment. The disadvantage is that due to the small infrared penetration ability, there are some limitations, most of the non-conductive material penetration thickness is less than 1 mm. Therefore, infrared rays can only detect the characteristic thermal electric field distribution formed on the surface of the equipment, and cannot detect the internal operation state from the outside of the equipment.

The common faults of high voltage equipment detected by infrared imaging method are as shown in Table 4-4.

Table 4-4 Common Faults Detected by Infrared Imaging in High-Voltage Equipment

Equipment type	Fault Type	Fault Characteristics
Transformer	Poor internal or external wiring, broken or loose leads, partial overheating of the iron core	partial overheating at the fault point
Instrument transformer	Internal connection faults, lack of oil	The overall increase in dielectric loss and temperature rise
Circuit breaker	Poor contact of line clamps, internal moisture, and overall insulation degradation	The large temperature difference between phases, uneven heating
Power capacitor	Moisture, loose connections, short circuit in windings, insulation aging, discharge from supports, lack of oil, poor impregnation	partial heating
Lightning arrester	Moisture, aging of valve plate resistance	Uneven longitudinal temperature distribution
Cable	Poor conductor connection, partial defects in cable heads, overall insulation deterioration, conductivity in gaps	Overheating at the defective part of the cable
High-voltage bushing	Poor sealing, poor contact of joints, oil leakage, moisture	Increase in dielectric loss and insulation contact resistance, leading to heating
Generator	Poor contact of joints, partial short circuit overheating of the iron core, defects in brushes and collector boots, magnetic leakage	Partial heating

Ultraviolet imaging method is mainly based on the theory of corona discharge. When high-voltage equipment experiences corona discharge, it emits ultraviolet radiation. After being received and processed by an ultraviolet camera, the image overlaps with visible light images and is displayed on the screen. This allows for direct observation of surface discharge of high-voltage equipment, determining the location, shape, and intensity of corona discharge. Therefore, it can detect insulation defects such as equipment damage, dirt, partial discharge caused by carbonization, erosion damage, and insulation degradation, accurately determining the health condition of the equipment.

The wavelength of infrared is in the range of 8–14 μm, while the wavelength range of ultraviolet is in the range of 40–400 nm. The difference between ultraviolet and infrared is their wavelength. Due to the ability to operate in daylight ("day-blind band"), ultraviolet imaging remains stable and clear even under bright sunlight. It is suitable for detecting the location and intensity of corona discharge, arcs, and surface partial discharge caused by cracks, dirt, surface damage, and looseness in power equipment. It is particularly effective in detecting corona discharge locations in insulators, wire connections, equipotential rings, and other power equipment in overhead lines and high-voltage substations. During measurement, attention should be paid to the influence of temperature, humidity, air pressure, wind speed, field of view angle, detection distance, instrument gain, and other factors on the accuracy of corona discharge detection to reduce measurement errors.

3. Leakage current of lightning arrester

Currently, non-gapped zinc oxide lightning arresters (Metal Oxide Arrester, MOA) are mainly used in substations, and dampness and aging are the main causes of lightning arrester failures. Under normal operating voltage, the current passing through the lightning arrester is very small, ranging from tens to hundreds of microamps (μA), which is called AC leakage current under operating voltage, also known as total current. Under normal circumstances, the capacitive component accounts for the majority of the total current, while the resistive component accounts for about 10% to 20% of the total current. When the lightning arrester ages or gets damp, its resistive current gradually increases, and the proportion of resistive current in the total current also increases accordingly, leading to an increase in the total current. Therefore, this characteristic can be used to judge the operating state of metal oxide.

The increase in resistive current reflects the defects of the lightning arrester when it is severely damp, aged, or has serious insulation degradation. By monitoring the total current or resistive current of the lightning arrester during operation and comparing the obtained data with the factory and historical data, insulation defects of the lightning arrester can be discovered. In the field, the third harmonic current method is used to obtain the total current from the lightning arrester grounding line, and the third harmonic is obtained through a third harmonic bandpass filter. By detecting the changes in harmonic current, the aging state of the metal oxide lightning arrester can be determined.

The initial current value of MOA refers to the current value passing through it measured during commissioning. By monitoring the AC leakage current and resistive component of the zinc oxide lightning arrester under operating voltage, insulation defects such as aging or dampness can be determined. The change in resistive current flowing through the zinc oxide lightning arrester under rated operating voltage directly reflects the quality of its insulation performance, with an increase in resistive leakage current being used as a judgment criterion. When the fundamental component of the resistive current increases significantly and the content of harmonics does not increase significantly, it generally indicates severe pollution or dampness. When the content of harmonics in the resistive current increases significantly and the fundamental component does not increase significantly, it generally indicates aging.

The main methods of MOA online monitoring are shown in Table 4-5.

Table 4-5 Main methods for online monitoring of MOA

Diagnostic Method	Description	Example
Full current method	The method is simple and feasible, sensitive to moisture detection, but not sensitive to early-stage aging. It often cannot be found until the resistive current component is increased several times	AC or Rectifier-type Ammeter
Resistive current component method	The method is more complex but can easily detect early-stage aging	Resistive Current Meter
Power loss	Measurement of the power loss caused by the resistive current component increase	Power Loss meter
Component temperature	Measurement of the temperature rise of the MOA caused by power loss	infrared camera

Measurements are taken under operating voltage to assess the full current and resistive current, reflecting the condition of the MOA. Changes in the full current can indicate severe moisture ingress, poor internal component contacts, and significant valve aging. However, changes in resistive current are more sensitive to early-stage valve aging. For example, when the peak resistive current increases from 50 μA to 250 μA, the increase in overall current may only be a few percentage points.

To measure the full current, a continuous online monitoring device is connected in series between the bottom of the lightning arrester and the ground to monitor the full current. When measuring the resistive current, the stability of the valve plate of the lightning arrester mainly relies on monitoring the resistive current. The resistive current is more sensitive to early-stage valve aging and moisture ingress. Since the current values of the arrester under operating voltage are small (in the μA range), strict requirements are placed on the sensor. When selecting a leakage current sensor for the arrester, high sensitivity and good stability sensors are chosen, taking into account the effects of interphase interference while monitoring the resistive current. During measurements, attention should be given to weather conditions, voltage fluctuations, wiring methods, electromagnetic interference, and stray currents, among other factors that may affect the accuracy of the measurements.

The influence of humidity can cause an increase in the external insulation leakage current of the arrester, leading to significant changes in the online measurement values. Humidity is also essential data when conducting longitudinal comparisons of arresters. During online monitoring of temperature and humidity, an external temperature and humidity meter installed on-site is used to collect temperature and humidity information, which is then transmitted to the online monitoring system through data conversion devices. The charged tester for zinc oxide arresters (see Table 4-4) can be used in places where the arrester is energized, de-energized, or in a laboratory, enabling simultaneous measurement of three-phase current and voltage. Wiring diagram for AC leakage current test under operating voltage is shown in Figure 4-15. Data analysis of zinc oxide arrester resistive current live detection is shown in Table 4-6.

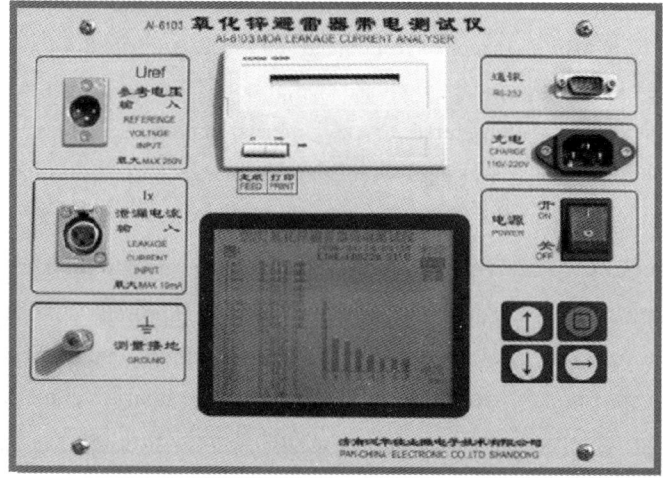

Figure 4-14 Zinc oxide arrester live tester

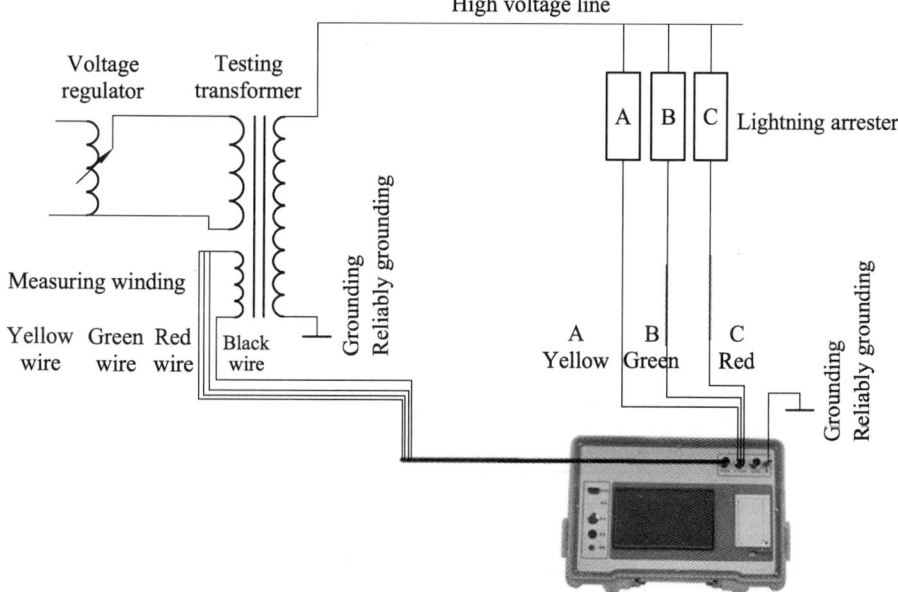

Figure 4-15 Wiring diagram for AC leakage current test under operating voltage

Table 4-6 Data analysis of zinc oxide arrester resistive current live detection

Observation indicator	Value	Analysis
Ratio of resistive current to full current	0%–20%	The lightning arrester is in good condition
	25%–40%	Strengthen monitoring and pay attention to changes in trends
	40% or more	Exit running and investigate the cause of the fault

There should be no significant changes in the full current, resistive current, and initial values. If the resistive current doubles, the power should be turned off for inspection. If the resistive current increases to 1.5 times its initial value, monitoring should be strengthened.

If the percentage of resistive current to full current increases significantly, with a large increase in the fundamental wave and no significant increase in harmonic waves, it can be determined that the lightning arrester is severely contaminated or has internal moisture ingress. If the percentage of resistive current to full current increases significantly, with a larger increase in harmonic waves and no significant increase in the fundamental wave, it can be determined that the arrester is aging.

4. Live Detection of Capacitive equipment capacitance and Dielectric Loss

Capacitance monitoring of capacitance equipment is applicable to cable systems, tap grounding end, current transformers, capacitive voltage transformers, capacitive bushings, capacitive couplers, and other capacitance-type devices. It utilizes a comparative method to measure the dielectric ratio in order to detect internal conditions such as moisture ingress in the equipment.

In the field, a 0.1 Hz ultra-low frequency dielectric loss test is commonly used to determine if electrical equipment is affected by moisture or aging. During the test, a 0.1 Hz power supply is applied to apply 0.5, 1.0 and 1.5 times the rated phase voltage U_0 to the power equipment. Multiple measurements of the dielectric loss are taken at each voltage level to obtain the average value and standard deviation for assessing the health of the electrical equipment.

5. Live detection of oil chromatographic

The principle of oil chromatographic live detection system is through the live detection of gas H_2, CO, CH_4, C_2H_6, C_2H_4, and C_2H_2 dissolved in transformer insulating oil, the type and concentration of transformer oil gas are detected to determine whether there is a fault or latent problem in the transformer, as shown in Table 4-7. The live detection process of oil chromatography is that the transformer oil enters the oil and gas separation chamber for oil and gas separation. The separated gas flows through the chromatographic column to separate the fault gas in turn.

Table 4-7 Transformer Faults and Increased Gas Components Table

Fault type	Primary Increased Gas Components	Secondary Increased Gas Components
Oil overheating	CH_4, C_2H_4	H_2, C_2H_6
Oil and paper overheating	CH_4, C_2H_4, CO, CO_2	H_2, C_2H_6
Partial discharge in oil-paper insulation	H_2, CH_4, C_2H_2, CO	C_2H_6, CO_2
Spark discharge in oil	C_2H_2, H_2	
Arc discharge in oil	C_2H_2, H_2	CH_4, C_2H_4, C_2H_6
Arc discharge in oil-paper insulation	C_2H_2, H_2, CO, CO_2	CH_4, C_2H_4, C_2H_6
Moisture or air bubbles in oil	H_2	

The gas sensor monitors the gas and converts the fault gas and gas concentration into electrical signals. The electrical signal is processed, and the type and concentration of each fault gas are calculated. The data is recorded and saved, and the critical value of the fault gas concentration is set in the system. When the set value is exceeded, the system will alarm to realize the live detection of the transformer fault.

To a certain extent, online monitoring replaces some conventional power outage preventive tests. However, because online monitoring cannot measure the AC parameters of power equipment above the operating voltage, it cannot obtain the insulation characteristics under DC voltage. Therefore, centralized real-time monitoring and portable online monitoring are now used, combined with conventional preventive tests when power equipment is shut down.

Module 3 Workshop

Ⅰ. Transformer Oil Chromatographic Online Monitoring System (Using Transformer Oil Chromatographic Online Monitoring as an Example)

The interface of the transformer oil chromatographic online monitoring system is shown in Figure 4-16.

Figure 4-16 Interface of transformer oil chromatographic online monitoring system

The transformer oil chromatographic online monitoring system consists of the following three parts, as shown in Figure 4-17.

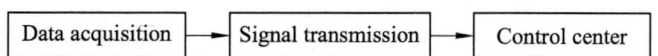

Figure 4-17 Composition of high-voltage online monitoring system

1. Data Acquisition

Principle: By monitoring the dissolved gas H_2, CO, CH_4, C_2H_6, C_2H_4 and C_2H_2 in transformer insulation oil online, the type and concentration of gas in the transformer oil can be detected, thereby determining whether the transformer has faults or potential problems.

Process: The transformer oil enters the oil and gas separation chamber for separation. The separated gas passes through the chromatographic column to sequentially separate the fault gas. The gas sensor monitors the gas to convert the concentrations of all fault gases and gases into electrical signals.

2. Information Transmission

The electric signals collected and processed are transmitted to the data control and processing center through the transmission device.

3. Control Center

Data processing is carried out on the electric signals to calculate the types and concentrations of various fault gases. The data is recorded and saved. The system sets critical values for the concentration of fault gases. When the set value is exceeded, the system will sound an alarm, realizing online monitoring of transformer faults.

II. Use of Transformer Oil Chromatographic Online Monitoring System

The transformer oil chromatographic online monitoring system has the following functions, which need to be learned, mastered, and trained, as shown in Figure 4-18.

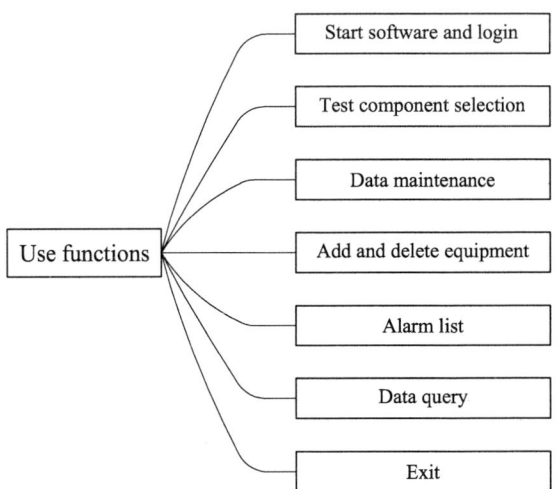

Figure 4-18 Functional composition of transformer oil chromatographic online monitoring system

1. Software Startup and Login

Entering a username and password gives more permissions, such as data maintenance, adding and deleting equipment, etc.

2. Component Selection for Detection

Select the gases that need to be detected from the components, as shown in Figure 4-19.

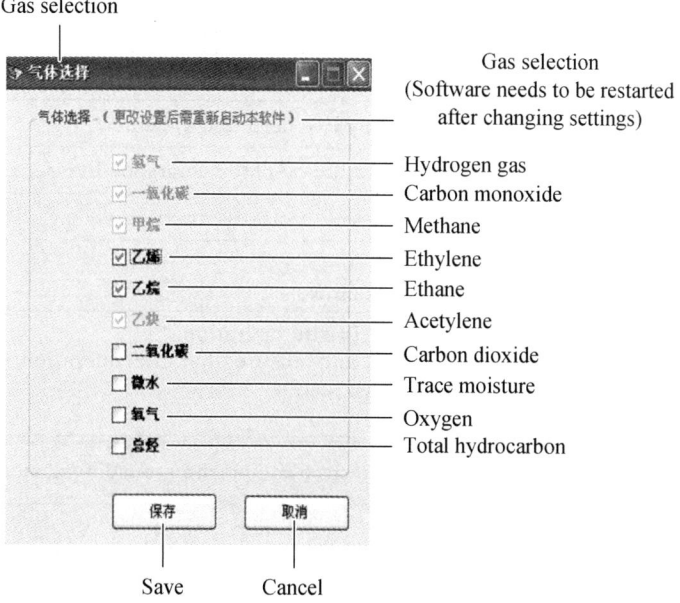

Figure 4-19 Detection component selection interface

3. Data Maintenance

Implement data backup and restore.

4. Add and Delete Devices

Add or delete transformers that need to be tested.

5. Alarm List

Query the alarm status of daily test data.

6. Data Query

Query historical test data and overall reports.

7. Exit

Close and exit the system.

Module 4   Training

What are the main characteristics of live detection and diagnosis technology?

Module 5 Evaluation

Practical test, describing the key points of the test, etc. As shown in Table 4-7.

Table 4-7 Evaluation form

Project Name	Live detection of transformer oil chromatography	Appraisal
Test equipment	Transformer oil chromatography live detection system	
Test content	Functional operation of transformer oil chromatography live detection	
Safety tools	Insulating shoes, insulating gloves, protective barriers, signage	
Potential risks	(1) Equipment damage. (2) Electrocution injuries or fatalities	
Project requirements	(1) On-site demonstration of in-situ operation. (2) Pay attention to safety and ensure that the operation process complies with safety regulations. (3) Writing the test report	
Material Preparation	Test instruments, personal protective equipment, grounding devices, etc.	
Safety risk control	(1) Before the test do "two-wearing and three-carrying" (wear work clothes, wear insulating boots, wear a helmet, wear insulating gloves, and use a voltage tester). (2) Set up barriers at the test site and hang "No Entry, High Voltage Danger" signs facing outward.	
Test Wiring	Wiring according to equipment requirements, with reliable grounding wire	
Test process	Skilled operation	
Data recording	Record test data, compare last test results	
Site cleanup	Exit the live detection system and remove the protective barriers and signage	
Result analysis	Determine equipment insulation status based on relevant standards	

Appendix

Training answer

Project 1 Overview of High Voltage Testing

Worksheet 1 High Voltage Safety

1. Answer: ① When installing or removing the grounding wire, two people must work at the same time, the operator and the guardian must wear insulating boots and helmet, the operator wears insulating gloves.

② When checking electricity, an electroscope with appropriate and qualified voltage level must be used. Before checking electricity, the electroscope should be tested on the powered equipment to confirm that it is good, then check the electricity on the power-off equipment, and finally double check it on the electric equipment. While checking electricity, all the lead-in and lead-out lines of the equipment to be tested must be tested.

2. Answer: ① When installing or removing the grounding wire, two people must work at the same time, the operator and guardian must wear insulating boots and helmets, and the operator wears insulating gloves.

② Installation and removal of grounding wire should be supervised. Installed grounding wire should be good contact, reliable connection.

③ Installation of grounding wire should be connected to the grounding terminal firstly, then connected to the conductor terminal at last. The order of removing the grounding wire is the opposite.

Worksheet 2 High-voltage Insulation Tools

1. Answer: ① Regular inspection once a year, test, qualified before used.

② The working voltage must be the same as the voltage level of the measured electrical equipment or power line, and the effective length of the pole should meet the corresponding requirements according to different voltage levels.

③ During the operation, a special person must be monitored. During the operation, the hand must be placed under the insulation rod buckle ring and must not exceed the guard ring.

2. Answer: The high-voltage electroscope should be checked whether it is consistent with the voltage level of the tested equipment, whether the insulation is good, whether the acousto-optic electricity works normally through self-checking, whether the label and qualification certificate are complete, and whether the qualification certificate is within the validity period of the test.

Worksheet 3　High-voltage Testing and Instruments

1. Answer: ① Insulation test is divided into non-destructive test and destructive test according to the degree of danger to the insulation of electric equipment. Therefore, in the test, the non-destructive test is carried out first, and the destructive test can be carried out after the test is qualified. If it is not qualified, the insulation recovery treatment is carried out first, such as drying, surface cleaning and so on.

② The difference between the tests is: non-destructive test refers to the test voltage applied to the test specimen is lower than the rated voltage of the electrical equipment to measure the insulation characteristics of the equipment, so as to determine whether there are defects within the insulation, this method will not damage the insulation.

Destructive test refers to the withstand voltage test on the electrical equipment which is much higher than the test voltage under the normal operation of the electrical equipment, and evaluates the bearing capacity and insulation margin of the electrical equipment when it encounters overvoltage. If the insulation margin of the electrical equipment fails to meet the requirements stipulated in the technical standards, insulation breakdown will occur during the withstand voltage test, resulting in damage.

③ Non-destructive tests are insulation resistance and absorption ratio measurement, DC leakage current measurement, dielectric loss measurement and most of the characteristics of the test. Destructive tests are DC withstand voltage test, AC withstand voltage test, impact withstand voltage test.

2. Answer: ① According to the different tasks, high voltage test can be divided into factory test, commissioning test (overhaul test) and preventive test. The factory test refers to the test of the high voltage performance and index of the finished product by the manufacturer. Commissioning test means that when new equipment is delivered, the owner and the manufacturer jointly test the equipment before it is put into operation. Preventive testing refers to test items after commissioning, The test contents of the three are basically the same, but the technical standards implemented are different.

② The factory test refers to the inspection and testing of the test items and each product by the power equipment manufacturer according to the relevant standards and product technical conditions. The purpose of the test is to check the quality of product design, manufacturing and processing and to prevent unqualified products from leaving the factory. A complete and qualified factory test report will be issued after the factory test.

The commissioning test refers to the test carried out by the installation department and the maintenance department on the new equipment and overhaul equipment according to the relevant standards and product technical conditions or procedures. The commissioning test before the new equipment is put into use is used to check whether the product is defective and whether there is damage during transportation. Testing of equipment after overhaul to check whether the quality of overhaul is qualified.

Preventive test is a test carried out by the operating department and the test department during a period time after the equipment has been put into service. The purpose of this test is to check out whether there are any insulation defects in operation. Compared with factory test and commissioning test, it mainly focuses on insulation test and has fewer test items.

Project 2   High Voltage Insulation Test

Worksheet 1   Insulation Resistance Test

Answer: Measuring insulation resistance is highly sensitive in inspecting the overall insulation condition of transformers. It can effectively detect moisture, surface dirt, or puncture defects in the insulation. It can also effectively reveal overall insulation moisture or permeability defects such as various short circuits, grounding issues, or cracked porcelain components.

Worksheet 2   Leakage Current and DC Withstand Voltage Test

Answer: The DC withstand voltage test is effective in detecting overall insulation defects caused by moisture or dirt and can reveal local defects by examining the curve of voltage and leakage current. It is less destructive to insulation and requires equipment with smaller capacity, making it more portable.

The AC withstand voltage test is effective in identifying dangerous focal defects and is the most direct method for assessing the insulation strength of electrical equipment. It is crucial in determining whether electrical equipment can be put into operation and is an important means of maintaining insulation level and preventing insulation accidents.

Both DC and AC withstand voltage tests can effectively detect insulation defects, but each has its own characteristics and advantages. Therefore, the two methods cannot be used interchangeably. In some cases, both tests may be necessary to provide complementary results.

Worksheet 3   Dielectric Loss Test

Answer:
① When the voltage level of the transformer is 5 kV or above and the capacity is 8,000 kVA or above, the tangent of the dielectric loss angle ($\tan\delta$) should be measured.

② The tanδ value of the tested winding should not exceed 130% of the factory test value.

③ When the measured temperature does not match the factory test temperature, it should be converted to the same temperature for comparison.

Worksheet 4　AC Voltage Withstand and Series Resonance Withstand Test

Answer: The essence of AC withstand voltage test is to detect the insulation strength and quality of equipment under rated AC voltage. During the test, a certain AC voltage is applied to the specimen, and its insulation performance is judged by observing whether breakdown occurs. It is a crucial testing method that is widely used in various fields such as power industry, petrochemicals, rail transportation, aerospace, etc.

Worksheet 5　Partial Discharge Test

Answer: The problems that can be detected by partial discharge test are: ① defective joints; ② defective terminals; ③ contaminated or dirty terminals; ④ damaged cable insulation layer.

Partial discharge is a complex physical process. In addition to involving the transfer of charges and the dissipation of electrical energy, it also generates electromagnetic radiation, ultrasonic waves, light, heat, and new byproducts. From an electrical perspective, during discharge, there is charge exchange, electromagnetic wave radiation, and energy loss at the discharge site. The most obvious effect is reflected at the two ends of the test specimen where the voltage is applied, with the appearance of weak pulse voltages. When there is a discharge in the air gap of the test specimen, it is equivalent to the test specimen losing charge q, causing a sudden drop in its terminal voltage ΔU. This is generally a microvolt-level pulse superimposed on the kilovolt-level external applied voltage. The working principle of all partial discharge testing equipment is to detect these voltage pulses.

Project 3　High Voltage Characteristic Test

Worksheet 1　DC Resistance Test

1. Answer: DC resistance refers to the resistance exhibited by a component when a direct current passes through it. It represents the inherent and static resistance of the component.

The purpose of measuring DC resistance is to check the quality of electrical equipment windings or coils and the integrity of circuits, in order to identify issues such as wire breakage, loose joints, poor contacts, and inter-turn short circuits caused by mechanical stress during manufacturing or operation. Additionally, during temperature rise tests of transformer, the DC resistance values are used to calculate the corresponding temperature values under different loads.

Insulation resistance refers to the resistance corresponding to the leakage current flowing through the dielectric when a DC voltage is applied and the polarization process comes to an end after a certain period of time. Insulation resistance represents the resistance of the insulation layer in conductive equipment.

Measuring the insulation resistance of equipment is a convenient auxiliary method to assess its insulation status. In the field, megohmmeters are commonly used to measure insulation resistance. Since the voltage applied by the megohmmeter is lower than the working voltage of the device under test, this test is non-destructive, safe, and easy to perform. The measured insulation resistance value can reveal various conditions of the electrical equipment's insulation, such as low insulation resistance, insulation breakdown, and severe thermal aging.

In summary, DC resistance is used to measure the resistance of inductive loads, while insulation resistance is a specialized tool for assessing insulation strength. Generally, lower values of DC resistance are desirable, whereas higher values of insulation resistance are preferred.

2. Answer: Instrument transformer, generator, motor, shunt and wire cable, etc.

Worksheet 2　Loop Resistance Test

1. Answer: Testing the loop resistance can be used to inspect the welding quality of winding joints, short circuits, damage, incorrect lead wire connections, and the status of tap switches. It can effectively identify manufacturing defects such as improper coil selection, loose connections at connection points, missing strands, and broken wires.

2. Answer: Good contact of the conductive circuit of the circuit breaker is an important condition to ensure its safe operation. An increase in loop resistance will cause severe heating of the contact points, resulting in spring annealing and burnout of the insulation components around the contact points. Therefore, it is necessary to measure the DC resistance of the conductive circuit in preventive testing.

Worksheet 3　Transformer Ratio and Grouping Measurement

Answer: For the transformer with a voltage below 35 kV and a turns ratio less than 3, the allowed deviation for the turns ratio is ±1%. For all other transformers, the allowed deviation for the rated tap voltage ratio is ±0.5%, and the voltage ratios for other taps should be within 1/10 of the transformer impedance voltage value, but not exceeding ±1%.

Worksheet 4　Transformer Winding Deformation Test

1. Answer: The degree of deformation is determined based on the transformer winding deformation measurement results. The degree of deformation can be classified as normal, moderate deformation, or severe deformation. A normal degree of deformation indicates that the transformer is in its original state or has no significant deformation, and can continue to operate without the need for winding repair. Moderate deformation refers to a situation where the transformer has obvious deformation but can still operate normally. It requires increased

monitoring and appropriate maintenance scheduling, as any subsequent short circuits or impacts are likely to cause substantial damage to the transformer, necessitating winding repairs or replacement. Severe deformation refers to a scenario where the transformer cannot continue operating due to deformation, and immediate handling is necessary.

2. Answer: During the transformer DC resistance test, the transformer is charged and residual magnetism may be present. The transformer winding deformation test is conducted by sweeping frequency response from 20 Hz to 1,000 Hz at different frequencies, with high accuracy. It can be affected by the DC resistance test. Therefore, the winding deformation test should be done before the DC resistance test.

Worksheet 5   Transformer Oil Chromatographic Analysis

Answer: Transformer oil chromatographic analysis is a method that uses gas chromatography to determine the composition and content of dissolved gases in insulating oil. It is an effective means for power supply enterprises to assess the existence of potential faults such as overheating and discharges in operating oil-filled electrical equipment, in order to ensure the safe and efficient operation of the power grid. It is also a necessary method for manufacturers of oil-filled electrical equipment to conduct factory tests on their products.

Worksheet 6   Insulating Oil Dielectric Strength Test

Answer:
(1) Wipe the electrode surface and electrode rod repeatedly with a clean silk cloth.
(2) Adjust the electrode gap with a standard gauge.
(3) Rinse with anhydrous ethanol for 3–4 times, and then blow dry with a hairdryer. Then clean the test oil sample 2–3 times.

Worksheet 7   Switch Characteristic Test

Answer: The opening and closing speed, opening and closing time, synchronization degree of opening and closing at different intervals, and the operating voltage of the opening and closing coils of the high-voltage switch are tested.

Worksheet 8   Cable Fault Comprehensive Test

Answer: Cable fault finding generally consists of four steps: fault nature diagnosis, fault location, route tracing, and fault location.

Worksheet 9   Insulating shoe and insulating glove test

Answer: (1) Check for cracks, air leakage, and the presence of a certification of conformity on the appearance.

(2) Fill the gloves with water and place them in a container filled with the same amount of water. The water levels inside and outside the gloves should be the same, with 90 mm of the gloves exposed above the water surface while keeping them dry and clean.

(3) Connect one end of an iron chain to an electrode and place the other end inside the insulating gloves.

(4) Press the test button and sing the response.

(5) Gradually increase the voltage at a constant rate to the specified value while recording the current value. If the current value is lower than the specified limit, the test is considered passed.

Worksheet 10　High Voltage Nuclear Phase Test

Answer: (1) During the nuclear phase, two people must work together. The operator and the supervisor must both wear insulating boots and safety helmets, while the operator should wear insulating gloves.

(2) During testing, it is strictly forbidden to simultaneously hook two bare conductors, as it may cause a short circuit between the two bare conductors.

(3) For bare conductors below 35 kV, a contact-type nuclear phase should be used. For bare conductors above 35 kV, a non-contact nuclear phase should be used. Non-contact nuclear phase: When the voltage of the bare conductor exceeds 35 kV, a non-contact nuclear phase must be employed. Gradually bring the detector close to the tested conductor, and when an electric field signal is detected, the nuclear phase can be completed.

Worksheet 11　Grounding Resistance Measurement of Ground Grid

Answer:

Four-point method:

Advantages: It can effectively eliminate the mutual inductance effect on the voltage measurement leads, and through phase reversal, it can eliminate the influence of interference currents in the ground, thus obtaining the true grounding resistance value.

Disadvantages: When the soil is non-uniform, the resistivity measured by the four-point method is not the actual resistivity, but a comprehensive value considering the non-uniformity of the soil. If the ground resistance tester used cannot automatically eliminate the mutual inductance effect, the current lead and voltage lead should be kept sufficiently far apart to reduce the mutual inductance effect. In addition, the use of multiple electrodes increases the workload of on-site measurements.

Three-point method:

Advantages: The resistance of the potential electrode and current electrode can be larger than the resistance of the tested grounding electrode, without affecting the measurement accuracy in essence.

Project 4 High Voltage New Technology

Worksheet 1 Hot Washing

Answer: Hot washing is a method of cleaning electrical insulation components that are contaminated while high-voltage equipment is operating. This method involves using a specially designed pump and mechanical device to spray water with a resistivity of not less than 1.5 kΩ · cm onto the components while maintaining a certain water pressure and safe distance. The purpose of hot washing is to remove dirt from the insulating surfaces while ensuring safety.

Worksheet 2 High Voltage Live Detection

Answer: In the condition-based maintenance of electrical equipment, the characteristics of point detection and diagnosis technology are mainly manifested in the following aspects:

(1) It does not need a power outage in the process of detection, which avoids the economic losses caused by a power outage and improves the reliability of the power supply effectively.

(2) The diagnosis technology of live detection technology can find some problems that occur during the operation of the equipment, and some old equipment cannot withstand the instantaneous high voltage. Therefore, the power failure withstand voltage test cannot be carried out, and the live detection diagnosis technology can avoid this problem.

(3) The cycle of detection and diagnosis can be arranged flexibly so that the potential safety hazards in the equipment can be detected in time.

Bibliography

[1] 住房和城乡建设部. 电气装置安装工程电气设备交接试验标准：GB 50150—2016[S]. 北京：中国计划出版社，2016.

[2] 国家能源局. 电力设备预防性试验规程：DL/T 596—2021[S]. 北京：中国电力出版社，2021.

[3] 国家能源局. 接地装置特性参数测量导则：DL/T 475—2017[S]. 北京：中国电力出版社，2017.

[4] 国家能源局. 现场绝缘试验实施导则：DL/T 474.1～474.5—2018[S]. 北京：中国电力出版社，2019.

[5] 国家能源局. 输变电设备状态检修试验规程：DL/T 393—2021[S]. 北京：中国电力出版社，2022.

[6] 国家能源局. 电力安全工器具预防性试验规程：DL/T 1476—2015[S]. 北京：中国电力出版社，2015.

[7] 国家能源局. 带电作业工具、装置和设备预防性试验规程：DL/T 976—2017[S]. 北京：中国电力出版社，2018.

[8] 中国南方电网责任有限公司. 10 kV 配电线路带电作业指南[M]. 北京：中国电力出版社，2015.

[9] 吴广宁. 高电压技术[M]. 北京：机械工业出版社，2007.

[10] 王亚妮. 高速铁路变配电设备检修岗位[M]. 北京：中国铁道出版社，2012.

[11] 何发武. 城市轨道交通电气设备测试[M]. 成都：西南交通大学出版社，2017.

[12] 张国光. 电气设备带电检测技术[M]. 北京：中国电力出版社，2014.